Food Lipids

ACS SYMPOSIUM SERIES **920**

Food Lipids

Chemistry, Flavor, and Texture

Fereidoon Shahidi, Editor
Memorial University of Newfoundland

Hugo Weenen, Editor
Wageningen Centre for Food Sciences

Sponsored by the
**ACS Division of Agricultural and
Food Chemistry, Inc.**

American Chemical Society, Washington, DC

Library of Congress Cataloging-in-Publication Data

Food lipids : chemistry, flavor, and texture / Fereidoon Shahidi, editor, Hugo Weenen, editor ; sponsored by the ACS Division of Agricultural and Food Chemistry, Inc.

p. cm.—(ACS symposium series ; 920)

Includes bibliographical references and index.

ISBN-13: 978-0-8412-3896-1 (alk. paper)

1. Lipids—Congresses. 2. Oils and fats—Flavor and odor—Congresses.

I. Shahidi, Fereidoon, 1951– II. Weenen, Hugo, 1953– III. American Chemical Society. Division of Agricultural and Food Chemistry, Inc. IV. Series.

TP453.L56F66 2005
664—dc22
                             2005048307

The paper used in this publication meets the minimum requirements of American National Standard for Information Sciences—Permanence of Paper for Printed Library Materials, ANSI Z39.48–1984.

Copyright © 2006 American Chemical Society

Distributed by Oxford University Press

ISBN 10: 0-8412-3896-0

All Rights Reserved. Reprographic copying beyond that permitted by Sections 107 or 108 of the U.S. Copyright Act is allowed for internal use only, provided that a per-chapter fee of $30.00 plus $0.75 per page is paid to the Copyright Clearance Center, Inc., 222 Rosewood Drive, Danvers, MA 01923, USA. Republication or reproduction for sale of pages in this book is permitted only under license from ACS. Direct these and other permission requests to ACS Copyright Office, Publications Division, 1155 16th Street, N.W., Washington, DC 20036.

The citation of trade names and/or names of manufacturers in this publication is not to be construed as an endorsement or as approval by ACS of the commercial products or services referenced herein; nor should the mere reference herein to any drawing, specification, chemical process, or other data be regarded as a license or as a conveyance of any right or permission to the holder, reader, or any other person or corporation, to manufacture, reproduce, use, or sell any patented invention or copyrighted work that may in any way be related thereto. Registered names, trademarks, etc., used in this publication, even without specific indication thereof, are not to be considered unprotected by law.

PRINTED IN THE UNITED STATES OF AMERICA

Foreword

The ACS Symposium Series was first published in 1974 to provide a mechanism for publishing symposia quickly in book form. The purpose of the series is to publish timely, comprehensive books developed from ACS sponsored symposia based on current scientific research. Occasionally, books are developed from symposia sponsored by other organizations when the topic is of keen interest to the chemistry audience.

Before agreeing to publish a book, the proposed table of contents is reviewed for appropriate and comprehensive coverage and for interest to the audience. Some papers may be excluded to better focus the book; others may be added to provide comprehensiveness. When appropriate, overview or introductory chapters are added. Drafts of chapters are peer-reviewed prior to final acceptance or rejection, and manuscripts are prepared in camera-ready format.

As a rule, only original research papers and original review papers are included in the volumes. Verbatim reproductions of previously published papers are not accepted.

ACS Books Department

Contents

Preface..ix

Flavor

1. **Importance of Non-Triacylglycerols to Flavor Quality of Edible Oils**..3
   Fereidoon Shahidi, Fayez Hamam, and M. Ahmad Khan

2. **Formation of Odor-Active Carbonyls in Self-Assembly Structures of Phospholipids**..19
   I. Blank, J. Lin, M. E. Leser, and J. Löliger

3. **The Effects of Diet, Breed, and Age of Animal at Slaughter on the Volatile Compounds of Grilled Beef**..................35
   J. Stephen Elmore, Donald S. Mottram, Michael Enser, and Jeffrey D. Wood

4. **Flavor Release from French Fries**..49
   Wil A. M. van Loon, Jozef P. H. Linssen, Alexandra E. M. Boelrijk, Maurits J. M. Burgering, and Alphons G. J. Voragen

5. **Flavor Release from Food Emulsions Containing Different Fats**........61
   M. Fabre, P. Relkin, and E. Guichard

6. **Changes in Key Odorants of Sheep Meat Induced by Cooking**..........73
   Valerie Rota and Peter Schieberle

Texture

7. **Differential Retention of Emulsion Components in the Mouth after Swallowing: ATR FTIR Measurements of Oral Coatings**..........87
   Harmen de Jongh, Anke Janssen, and Hugo Weenen

8. **The Role of Fats in Friction and Lubrication**............................95
   J. F. Prinz, R. A. de Wijk, and H. Weenen

9. Prediction of Creamy Mouthfeel Based on Texture Attribute Ratings of Dairy Desserts..........105
   H. Weenen, R. H. Jellema, and R. A. de Wijk

10. Effects of Structure Breakdown on Creaminess in Semisolid Foods..........119
    Hugo Weenen

11. Chemistry and Rheology of Cheese..........133
    Michael H. Tunick and Diane L. Van Hekken

Flavor and Texture

12. How Lipids Influence Flavor Perception..........145
    Kris B. de Roos

13. Release of Flavor from Emulsions under Dynamic Sampling Conditions..........159
    R. S. T. Linforth and A. J. Taylor

14. Fat Reduction in Foods: Microstructure Control of Oral Texture, Taste, and Aroma in Reduced Oil Systems..........171
    G. J. van den Oever

15. Effect of Composition of Triacylglycerols on Aroma Volatility: Application to Commercial Fats..........191
    Natacha Roudnitzky, Gaëlle Rondaut, and Elisabeth Guichard

16. Fatty Acid and Volatile Flavor Profiles of Textured Partially Defatted Peanuts..........205
    Margaret J. Hinds, M. N. Riaz, D. Moe, and D. D. Scott

Author Index..........221

Subject Index..........223

Preface

Lipids are a condensed source of energy; therefore, decreasing the lipid content of foods has been one of the main strategies to combat obesity. Obesity is an increasingly serious problem, especially in the more affluent parts of the world. However, decreasing the lipid content of foods has had only limited success, mainly because of adverse sensory effects that make such foods less attractive. To develop more successful strategies to replace lipids with less energy dense alternatives, a good understanding of the sensory contribution of lipids is required.

In foods, lipids occur as emulsions or as free oil/fat dispersed in a solid matrix. Lipids contribute to both the flavor and texture of foods. The contribution of lipids to flavor is due to volatile oxidation products as well as to the taste of short-chain free fatty acids, and because the microstructure of lipid emulsions affect flavor release. Moreover in human's brain, taste, flavor, and texture are integrated in the orbitofrontal cortex to give rise to overall perception associated with a certain food, or its attractiveness to various degrees. In other words, to improve our understanding of the contribution of lipids to the sensory appreciation of foods, it is necessary to understand its effects on all sensory modalities (taste, flavor, texture, and other trigeminal sensations) and their interactions. This book focuses on the main sensory modalities that are affected by lipids and their interactions, namely flavor, texture and flavor–texture interactions.

To understand the relation between the composition of a food emulsion and its sensory perception, knowledge of the microstructure of the emulsion, its physicochemical properties, and what happens in the mouth while eating is essential. Although flavor release under static conditions is determined mainly by molecular diffusion, during eating flavor release is primarily controlled by the rate of surface renewal.

From a microstructural viewpoint, emulsion properties affecting sensory perception can be divided into factors associated with the discrete phase (phase volume and droplet size distribution), those associated with the adsorbed layer (composition and quantity) and those associated with the continuous phase (viscosity, solution composition, and properties). The partitioning of flavor-active components between the discrete and continuous phase of a food emulsion plays an important role in the perception of the overall sensory properties.

This book contains contributions on a variety of subjects representing a state-of-the-art knowledge on flavor and texture of lipid-containing foods. Of the six chapters in the "Flavor" section, four chapters focus on the contribution of volatiles to the overall flavor of various foods (beef, mutton, peanuts, and French fries), one on the contribution of non-triacylglycerols to the flavor quality of oils, and one on the formation of odor-active compounds generated upon heating of aqueous dispersions of phospholipids.

The five chapters in the "Texture" section include discussion of the formation of an oral coating when eating a lipid-containing emulsion, the role of fats in oral friction and lubrication, the nature of creaminess, and the chemistry and rheology of cheese.

The section on "Flavor and Texture" starts with a review on how lipids influence flavor perception, release of flavor from emulsions under dynamic sampling conditions, and microstructural control of oral texture, taste, and aroma in reduced oil systems. The effect of various triacylglycerols on flavor release is also presented.

We gratefully acknowledge the financial support of the American Chemical Society Division of Agricultural and Food Chemistry, Inc. (The United States) and the Wageningen Centre for Food Sciences (The Netherlands). We extend our appreciation to authors and reviewers for their contributions.

Hugo Weenen
Numico Research and Development
WTC Schiphol Airport, Tower E
Schiphol Boulevard 105
1118 BG Schiphol Airport
The Netherlands

Fereidoon Shahidi
Department of Biochemistry
Memorial University of Newfoundland
St. John's, Newfoundland A1B 3X9
Canada

Food Lipids

Flavor

Chapter 1

Importance of Non-Triacylglycerols to Flavor Quality of Edible Oils

Fereidoon Shahidi, Fayez Hamam, and M. Ahmad Khan

Department of Biochemistry, Memorial University of Newfoundland, St. John's, Newfoundland A1B 3X9, Canada

Edible oils from vegetable and animal sources are composed primarily of triacylglycerols and to a lesser extent minor components referred to as unsaponifiable matter. During processing of oils and as a result of reactions in production of structured and other novel lipids many of these components are removed and hence the relative stability of the preparation is compromised. Presence of chlorophyll in the oils, both endogenous and in the additives, resulted in enhanced production of odor-active aldehydes under fluorescent light. However, simultaneous removal of non-triacylglycerol components reduced the stability of oils under Schaal oven conditions. In enzymic acidolysis of algal oils with capric acid we found that the resultant oils were much less stable than their unaltered counterparts despite a decrease in the degree of unsaturation of the products. Experiments carried out in the absence of any enzyme showed that removal of endogenous antioxidants was indeed responsible for this phenomenon.

Fats and oils provide a concentrated source of energy and essential fatty acids as well as fat soluble vitamins and other minor components. Lipids also serve as an important constituent of cell walls. In foods, lipids provide flavor, texture and mouthfeel to products. In addition, fats and oils serve as a heating medium and are important in the generation of aroma, some of which arise from direct interaction of lipids and/or their degradation products with food constituents.

Oilseeds and tropical fruits are a major source of food lipids, in addition to those consumed from land-based and aquatic animals. The edible oils from oilseeds may be produced by pressing, solvent extraction or their combination. The seeds may first be subjected to a pretreatment heating in order to deactivate enzymes present. The oils after extraction are subsequently subjected to further processing steps of degumming, refining, bleaching, deodorization, and if necessary, stabilization.

Fats and oils from source materials are composed primarily of triacylyglycerols (TAG). In addition, phospholipids, glycolipids, waxes, wax esters, tocopherols and tocotrienols, other phenolics, carotenoids, sterols and chlorophylls, and hydrocarbons, among others, may be present as minor constituents and these are collectively referred to as unsaponifiable matter (*1*). During processing, storage and use, edible oils undergo chemical and physical changes. Often, process-induced changes of fats and oils are necessary to manifest specific characters of food, however, such changes should not exceed a desirable limit. Both TAG and minor constituents of the oil exert a profound influence on quality characteristics of the oils and hence their effect on health promotion and disease prevention.

The left over meal, following oil extration, may also serve as a source of phytochemicals. Obviously, hulls, might be included in the meal if the seeds are not dehulled prior to oil extraction. The importance of bioactives in processing by-products of oilseeds, both as deodorizer distillate and as meal components, may thus require attention.

Fats and oils from animal sources originate from depot fat, liver oil, as well as intramuscular lipids. In aquatic animal, the source of lipid depends on the species involved; lean white fish reserving lipids in their liver while fatty fish deposit the lipids in the intramuscular region and marine mammals store them as blubber (subcutaneous). The non-triacylglycerol components, similar to plant oils, may again contribute to different extent in the isolated lipids. While the usual level is less than 5%, shark liver oil may contain a substantial amount of non-triacylglycerols, mainly squalene.

The quality characteristics of edible oils, as noted earlier, depend on the composition of their fatty acids, positional distribution of fatty acids, non-triacylglycerol components, presence/absence of antioxidants, the system in which the oil is present such as bulk oil versus emulsion and low-moisture foods, as well as the storage conditions. Many of the non-triacylglycerol components of the oils are recovered during processing. This may result in destabilization of the oil and hence addition of antioxidants may be necessary in order to impart adequate stability to the oils. Furthermore, non-triacylglycerols separated during processing of edible oils may be recovered and used as important nutraceuticals intended for health promotion and disease prevention.

Triacylglycerols

Neutral lipids, mainly triacylglycerols, usually account for over 95% of edible oils. The fatty acids present are either saturated or unsaturated (Figure 1). Although saturated lipids are generally condemned because of their perceived negative effect on cardiovascular disease, recent studies have shown that C18:0 is fairly benign while C14:0 may possess adverse health effects. Furthermore, a better understanding of detrimental effects of trans fats has served as a catalyst for the return from using hydrogenated fats in place of naturally-occurring edible oils with a high degree of saturation. Hence, palm oil now serves as an important constituent of non-hydrogenated margarine formulations.

Of the unsaturated lipids, monounsaturated fatty acids, similar to their saturated counterparts, are non-essential as they could be synthesized in the body *de novo*. However, polyunsaturated fatty acids (PUFA; containing 2 or more double bonds) could not be made in the body and must be acquired through dietary sources. The parent compounds in this group are linoleic acid (LA, C18:2ω6) and linolenic acid (LNA, C18:3ω3). The symbols ω6 (or n-6) and ω3 (or n-3) refer to the position of the first double bond from the methyl end group of fatty acids involved as this dictates the biological activity of the molecules concerned (2).

Dietary lipids are composed mainly (80%) of C18 fatty acids and it is recommended that the ratio of ω6 to ω3 fatty acids be at least 5:1 to 10:1, but the western diet has a ratio of 20:1 or less. Enzymes in our body convert both groups of PUFA through a series of desaturation and elongation steps to C20 and C22 products, some of which are quite important for health and general well-being. The C20 compounds may subsequently produce a series of hormone-like molecules known as eicosanoids which are essential for maintenance of health. Obviously, elongation of LA and alpha-linolenic acid (ALA) to other fatty acids may be restricted (to about 5%) by rate determining steps and lack of the required enzymes in the body (See Figure 1). Thus, pre-term infants and the elderly may not be able to effectively make even these limited transformations and that production of docosahexaenoic acid (DHA) from eicosapentaenoic acid (EPA) is rather inefficient. Furthermore, supplementation with gamma-linolenic acid (GLA) as a precursor to arachidonic acid (AA) might also be necessary.

Lipids are generally highly stable in their natural environment; even the most unsaturated lipids from oilseeds are resistant to oxidative deterioration prior to extraction and processing. Presence of endogenous antioxidants is responsible for adequate oxidative stability of the oils. Nature appears to be able to protect itself as higher level of antioxidants are generally found in highly unsaturated oils. It is

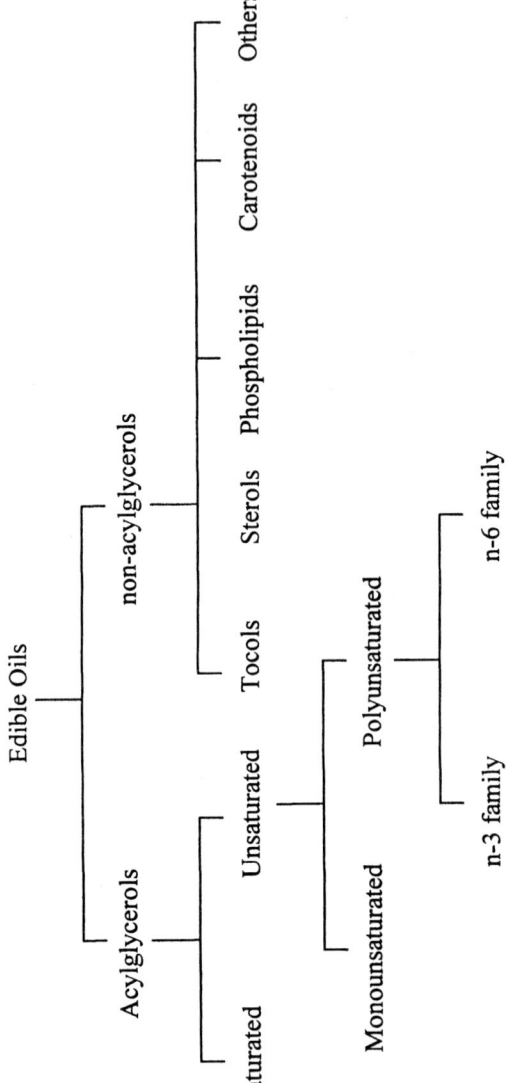

Figure 1. Edible oils and their fatty acid and non-triacylglycerol components.

believe that unsaturated oils generally co-exist with antioxidants in order to protect themselves from oxidation; of course the natural capsule or seed coat provides a barrier to light and oxygen as well as compartmentalization of oil cells and inactivity of enzymes prior to crushing. However, upon crushing and oil extraction, the stability of edible oils is compromised and this is dictated by several factors, including the degree of unsaturation of fatty acid constituents. This topic will be discussed in a later section. In addition, the position of fatty acids in the triacylglycerol molecule (Sn-1, Sn-2 and Sn-3) would have a considerable effect on their stability as well as assimilation into the body. Generally, fatty acids in the Sn-1 or Sn-3 position are hydrolyzed by pancreatic lipase and absorbed while those in the Sn-2 position are used for synthesis of new TAG and these might be deposited in the body.

Stability of Edible Oils

Stability of edible oils is affected by their constituent fatty acids and minor components, both endogenous and those added intentinally, as well as storage conditions. Thus as the degree of unsaturation of an oil increases, its susceptibility to oxidation, under similar conditions, increases. Furthermore, the condition in which the oil is present, e.g. bulk versus emulsion, has a profound effect on the stability of the oil. In addition, presence of antioxidants as well as metal ions, light, chlorophylls and other pigments might influence the stability of the oil. As an example, we found that while crude green tea extracts behaved pro-oxidatively in a bulk oil model system, removal of the chlorophyll reversed the situation (3).

Degree of Unsaturation

As the number of double bonds in a fatty acid increases, its rate of oxidation increases. Thus, docosahexaenoic acid (DHA, $C22:6\omega3$) is more prone to oxidation than eicosapentaenoic acid (EPA, $C20:5\omega3$) followed by arachidonic acid (ARA $C20:4\omega6$), alpha-linolenic acid (ALA, $C18:3\omega3$), and gamma-linoleic acid (GLA, $C18:3\omega6$), linoleic acid (LA, $C18:2\omega6$) and oleic acid (OA, $C18:1\omega9$). This trend is also reflected in the triacylyglycerols of different oils. Thus, as the iodine value (IV) of oils increases, their oxidation potential is also increased. Table I summarizes the induction period of several oils with different degrees of unsaturation as obtained using a Ranciment apparatus.

Table I. Rancimat induction periods of selected oils at 100°C

Oil	Induction period (h)
Palm	35.55
Palm (stripped)	17.38
Corn	20.79
Canola	15.59
Rice bran	12.07
Algal (DHASCO)	0.67
Seal blubber	0.17
Menhaden	0.13

Non-triacylyglycerols Constituents

The non-triacylyglycerol (NTAG) or unsaponifiable matter content varies from one oil to another. It appears that as the iodine value of oils increases, their content of tocols also increases (see Table II). However, this simple trend is not extended to fats and oils from animal sources. Different classes of compounds belonging to non-triacylglycerol constituents and their application areas are summarized in Table III. In most oils these constitute approximately 1-2%, but in others they may be present at 2-8% and in some cases even higher. Many of the unsaponifiable matter are recovered from oil during processing steps of degumming, refining, bleaching and deodorization. Thus, loss of phospholipids, sterols, tocopherols, carotenoids and related compounds during oil processing may range from 35 to 95%. These material may be collected as distillates during the deodorization process. Distillates that are rich in certain components may be separated and marketed for use in nutraceutical applications. Thus, mixed tocopherols, tocotrienols, carotenoids, lecithin and other constituents may be separated from soybean oil, palm oil, rice bran oil and barley oil, depending on their prevalence.

In a recent experiment on acidolysis of algal oils with capric acid we found that the stability of the resultant structured lipids was even lower than that of the original oils. This was despite the fact that products so obtained were more saturated. The non-triacylglycerol or unsaponifiable matter were implicated, as will be discussed under the effect of processing.

Table II. Correspondence of degree of unsaturated and tocol content of selected oils

Oil	Iodine Value (g/100g oil)	Tocol content (mg/100g)
Coconut	9	2
Illipe	33	10
Mango	47	15
Tea seed	86	20
Rapeseed	107	65
Soybean	135	100

Tocols and Ubiquinones

Tocols include both tocopherols (T) and tocotrienols (T3) which are present in edible oils in different compositions and proportions. Eight different compounds exist; each series of T and T3 includes four components designated as α, β, γ and δ, depending on the number and position of methyl groups on the chromane ring. The α-isomer is 5, 7, 8-trimethyl; β-isomer, 5, 8-dimethyl; γ-isomer, 7,8-dimethyl; and δ is the 8-methyl isomer.

The occurrence of tocopherols in vegetable oils is diverse, but animal fats generally contain only α-tocopherol. However, absence of α-tocopherol in blood plasma and other body organs may not negate the importance of other tocopherols such as γ-tocopherol. In oilseed lipids, there appears to be a direct relationship between the degree of unsaturation as reflected in the iodine value (IV), and the total content of tocols, mainly tocophenols (see Table II). Most vegetable oils contain α-, γ- and δ-tocopherols, while β-tocopherol is less prevalent, except for wheat germ oil. Meanwhile, tocotrienols are present, in large amounts, mainly in palm and rice bran oils. The antioxidant activity of tocotrienols generally exceeds that of their corresponding tocopherols. Meanwhile, the antioxidant activity of tocopherols is generally in the order of δ > γ > β > α; this is opposite to the trend for their vitamin E activity. With respect to ubiquinones, also known as coenzyme Q, they occur as 6 to 10 isoprene unit compounds; that is Q_6 (UQ-6) to Q_{10} (UQ-10). Coenzymes Q_{10} (UQ-10) and to a lesser extent Q_9 (UQ-9) are found in vegetable oils. Ubiquinone provides efficient protection *in-vivo* for mitochondria against oxidation, similar to vitamin E in the lipids and lipoproteins. The importance of coenzyme Q_{10} in cardiovascular health has been documented.

Table III. Non-triacylglycerols of edible oils and their use

Constituent	Example	Source	Application
Hydrocarbon	Squalene	Shark liver oil, olive oil, etc.	Skin care
Phospholipids	Phosphatidylcholine & Phosphatidylserine (Lecithin)	Vegetable oils	Dietary supplement, OTC
Sterols	β-Sitosterol	Canola oil, etc.	Functional food ingredient, OTC
Tocols	Tocopherols & tocotrienols	Palm oil, rice bran oil, etc.	Dietary supplement, OTC
Ubiquinone	Coenzyme Q10	Vegetable oils, etc.	Dietary supplement, OTC
Carotenoids	β-Carotene, xanthophylls	Palm oil, etc.	Dietary supplement, OTC
Phenolics	Phenolic acids, flavonoids, isoflavonoids	Oilseed meals, olive oil, etc.	Dietary supplement

Phospholipids

Phospholipids possess fatty acids which are generally more unsaturated than their associated tricylglycerols and hence are more prone to oxidation. However, this situation does not necessarily hold for marine oils, such as seal blubber oil, whose phospholipids are less unsaturated than its triacylglycerols. The role of phospholipids as pro- or antioxidants is, however, complex because in addition to their lipid moiety, they contain phosphorous- and nitrogen-containing groups that dictate their overall effect in food systems. The lecithins recovered from soybean oil are sold as nutraceuticals. Phosphatidylserine, one of the phospholipids in soybean and other oils has been found to enhance the memory and cognition.

Extensive studies have demonstrated that phospholipids may exert antioxidant effects in vegetable oils and animal fats. The exact mechanism of action of phospholipids in stabilizing fats and oils remains speculative; however, evidence points out to the possibility of their synergisms with tocopherols, chelation of prooxidant metal ions as well as their role in the formation of Maillard-type reaction products. King *et al.* (4) found a positive relationship between the presence and type of phospholipids and stability of salmon oil in the order given below.

Sphingomyelin ≈ lysophosphatidylcholine ≈ phosphatidylcholine ≈ phosphatidyl-ethanolamine > phosphatidylserine > phosphatidylinositol > phosphatidylglycerol.

Phytosterols

Edible oils generally contain a variety of sterols which exist in the free form, as sterol ester of fatty acids and sterol glycosides or esters of sterol glycosides. Sterols are heat-stable molecules with no flavor of their own and exhibit antipolymerization activity during frying. Sterols serve as a means for fingerprinting of vegetable oils and lend themselves for detection of adultration of oils. Among sterols, Δ^5-avenasterol, fucosterol and citrostadienol have been shown to exhibit antioxidant properties. Donation of a hydrogen atom from the allylic methyl group in the side chain is contemplated. Most oils contain 100-800 mg/100g sterols. Brassicasterol is specifically found in canola, rapeseed and mustard oils. Phytosterols may also be obtained from non-oil sources for addition to margarine, and, in some cases, to liquid oils, in order to take advantage of their cholesterol lowering activity (up to 15%) in human subjects.

Carotenoids

Carotenoids are widespread in oilseeds, but are found in the highest amount in palm oil at 500-700 ppm levels. Both hydrocarbon-type carotenoids, namely α-and β-carotene, as well as xanthophylls, oxygenated carotenes, are present. Carotenoids act as scavengers of singlet oxygen and hence are important in stability of oils exposed to light. While β-carotene, and α-carotene, are usually the dominant components, α-xanthophylls may be present in smaller amounts. The role of certain xanthophylls, such as lutein and zeaxanthin, as nutraceuticals has been acknowledged.

Chlorophylls

Chlorophylls are present in a variety of oils. In particular, extra-virgin olive oil contains a large amount of chlorophylls and this is often associated with the high quality of this oil. However, in oils such as canola, immature seeds contain chlorophylls which end up in the oil and affect its stability. Meanwhile, grapeseed oil is generally green in color. Due to their photosensitizing effects, chlorophylls lead to oxidative deterioration of oils when exposed to light.

Phenolics

Phenolic compounds, other than tocols, may be present in edible oils. While tocols are always present in the free form, phenolic acids, phenylpropanoids and flavonoids and related compounds occur in the free, esterified and glycosylated forms. These compounds reside mainly in the meal, but their presence in oils, such as olive oil, is well recognized. Olive oil contains a number of phenolics, including hydroxytyrosol. Furthermore, sesame oil contains sesamin, sesaminol and sesamol which render stability to the oil. Meanwhile, oat oil contains a number of ferulates that affect its stability. Extraction of phenolics from oilseeds, as well as other seeds, has been practiced. Phenolics isolated from such sources may be used as nutraceuticals in the tablet or capsule forms.

Hydrocarbons

Hydrocarbons are another group of unsaponifiable matter that may occur in edible oils. This class of compound includes squalene which constitutes up to one third of the unsaponifiables in olive oil and is also a main constituent of shark liver oil. The effect of squalene in stabilization of oils at high temperatures and in the body for protection of skin is recognized. However, in an oil system, under Schaal oven conditions, we did not find it to be antioxidative in nature.

Effect of Processing on Non-Triacylyglycerols Components and Stability of Oils

As explained earlier, during different stages of degumming, refining, bleaching and deodorization, 35-95% of non-triacylyglycerol components of edible oils may be removed (*5*). The effects of processing, as determined by Ferrari *et al.* (*6*), on the contents of tocopherols and sterols for canola, soybean and corn oils are summarized in Table IV. In addition, carotenoids in edible oils, especially in palm oil, might be depleted during processing. Bleaching of carotenoids might be carried out intentionally in order to remove the red color, however, carotenoids might be retained using cold pressing at temperatures of as low as 50°C.

In order to measure the effect of processing and contribution of minor components to the stability of edible oils, it is possible to strip the oil from its non-triacylyglycerol constituents. To achieve this, the oil is subjected to a multi-layered column separation. The procedure developed by Lampi *et al.* (*7*) may be employed. We used a column packed with activated silicic acid (bottom layer, 40 g) followed by a mixture of Celite 545 / activated charcoal (20 g, 1:2 (w/w)), a mixture of Celite 545/powdered sugar (80 g, 1:2 (w/w)) and activated silicic acid (40g) as the top layer. Oil was diluted with an equal volumn of n-hexane and passed through the column that was attached to a water pump; the solvent was then removed. The characteristics of the oils before and after stripping indicated that while oxidative products and most of the minor components were removed, γ-tocopherol was somewhat retained in the oil.

The oils (olive, borage and evening primrose) were subjected to accelerated oxidation under Schaal oven condition at 60°C or under fluorescent lighting. Results indicated that oils were more stable, as such, than their stripped counterparts when subjected to heating, but under light, the oxidative stability of oils stripped of their minor components was higher. Thus, evening primrose oil as such (non-stripped) was more stable than its stripped counterpart under autoxidative conditions (Figure 2). However, the reverse was true under photo-oxidation (Figure 3). Examination of the spectral characteristics of oils indicated that chlorophylls were present in the original oils and hence might have acted as photosensitizers leading to enhanced oxidation of unstripped oils (see Figure 4; *8,9*). Photosensitizing effect of chlorophylls was also confirmed when crude green tea extracts were added to marine oils. Decholorphillization, however, allowed antioxidant catechins to exhibit their known protective effects (*3*). In case of red palm oil, however, removal of carotenoids by stripping

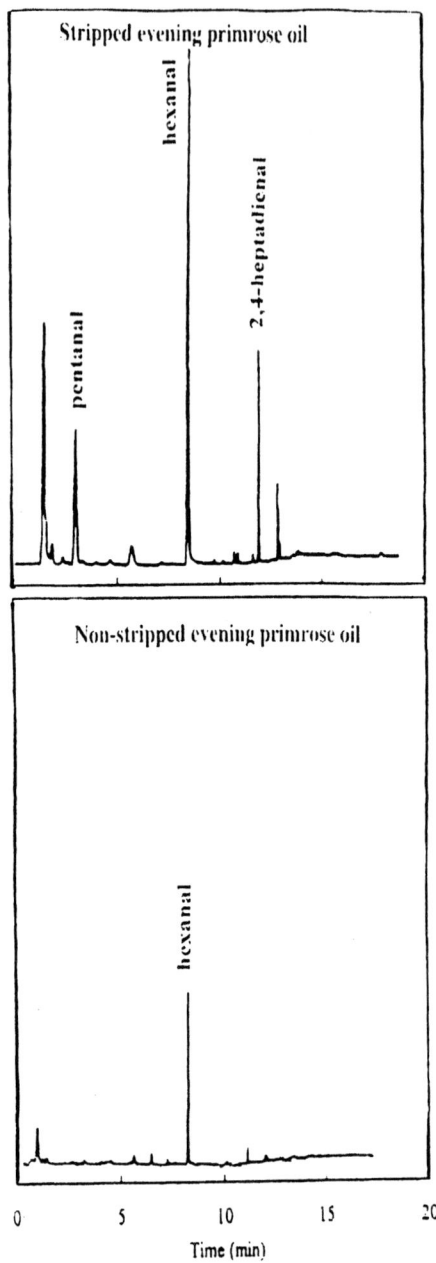

Figure 2. Stability of stripped and non-stripped evening primrose oil under Schaal oven condition at 60°C.

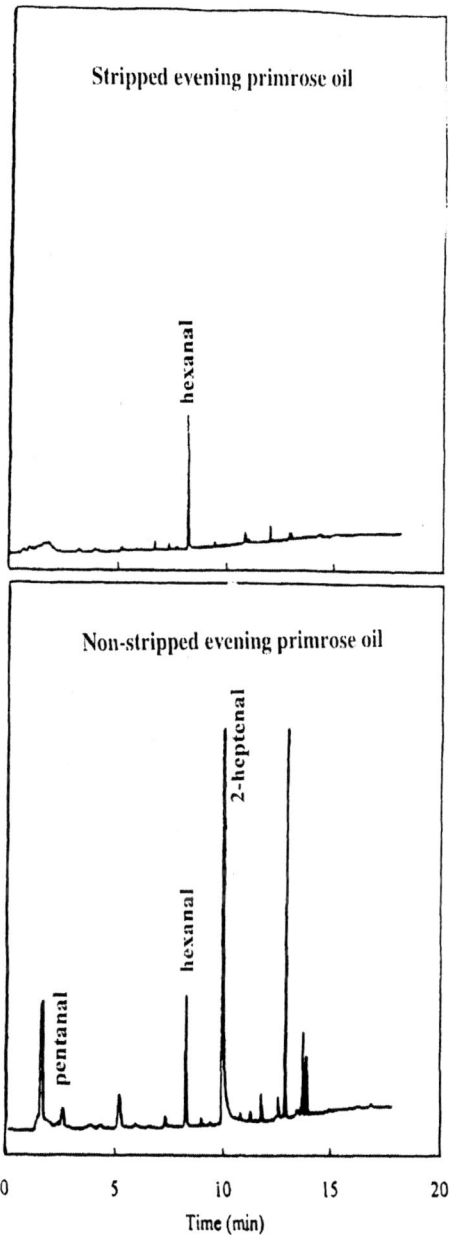

Figure 3. Stability of stripped and non-stripped evening primrose oil under Photooxidative conditions.

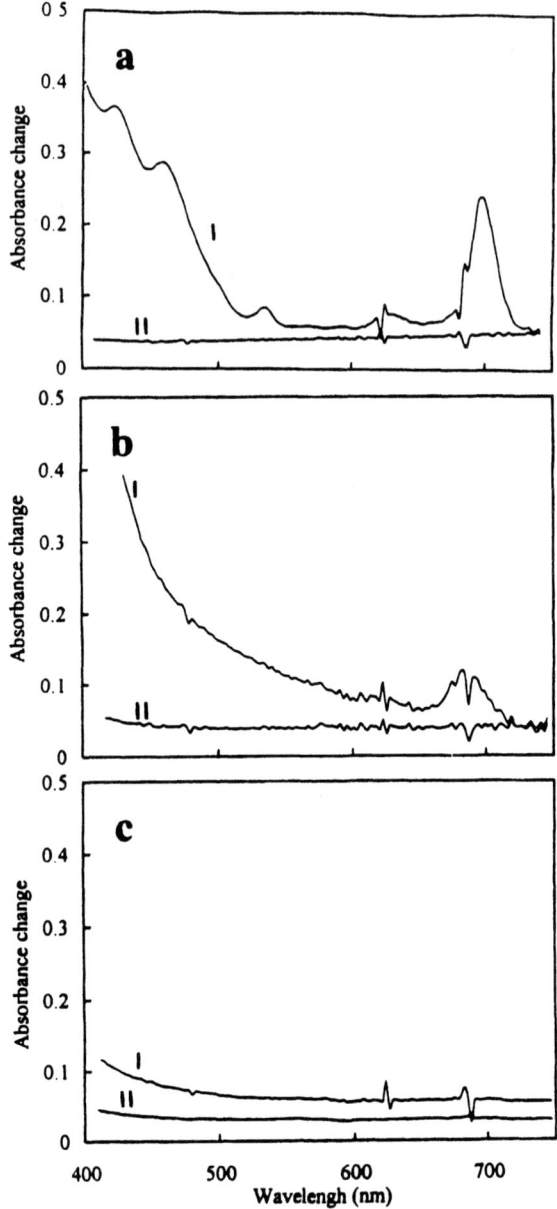

Figure 4. Removal of pigments, including chloropohylls, from olive (a), evening primrose (b) and borage (c) oils upon stripping process. Symbols I and II denote non-stripped and stripped oils, respectively.

Table IV. Changes in the content and composition of minor components of edible oils during processing (mg/100g).

Constituent	Crude	Refined	Bleached	Deodorized
Canola (Rapeseed)				
Tocopherols	136.0	128.7	117.8	87.3
Sterols	820.6	797.8	650.4	393.0
Soybean				
Tocopherols	222.3	267.7	284.0	195.2
Sterols	359.5	313.9	288.8	295.4
Corn				
Tocopherols	194.6	203.8	201.9	76.7
Tocotrienols	7.9	10.2	10.0	6.1
Sterols	1113.9	859.2	848.8	715.3

resulted in a decrease of near 15 h in its induction period as measured by Rancimat at 100°C (*10*). Therefore, minor components of edible oils and the nature of chemicals involved have a major influence on the stability of products during storage and food preparation.

Table V. Oxidative state of algal oils before and after modification.[a]

Sample	Unmodified Oil	Modified Oil	Processed Oil
DHASCO			
CD	9.4	30.8	59.0
TBARS	4.7	16.5	16.3
ARASCO			
CD	22.1	51.1	46.1
TBARS	14.7	13.9	14.2
OMEGA-GOLD			
CD	22.4	45.7	67.4
TBARS	11.1	11.7	17.5

[a]Modified oils contain different levels of capric acid; processed oil is the unmodified oil subjected to process steps in the absence of any enzyme. Abbreviations are: CD, conjugated dienes; TBARS, 2-thiobarbituric acid reactive substances. Values provided are after 48 h of storage at 60°C under Schaal oven conditions.

The processing steps in the production of specialty oils, such as structured lipids, may also lead to the removal of non-triacylglycerol components from the reaction medium and hence the isolated product. Table V shows that the stability of DHASCO (docosahexaenoic acid single cell oil), ARASCO (arachidonic acid single cell oil), and Omega-Gold oil was indeed better than that of their corresponding structured lipids containing capric acid. Removal of non-triacylglycerol components was suspected to be responsible for this observation. To test this assumption, experiments were carried out in which starting materials were subjected to the same reaction conditions in the absence of enzymes. The stability of DHASCO and ARASCO samples, subsequent to this operation, was indeed compromised, in line with exception (*11-13*). Similar results were observed for a third algal oil, omega gold, which also contained DHA as well as DPA (n-6 docosapentaenoic acid). Thus, importance of non-triacyglycerol components in stability and stabilization of fats and oils may exceed the effects exerted by the degree of unsaturation of fatty acids involved.

References

1. Shahidi, F.; Shukla, V. K. S. *INFORM* **1996**, *8*, 1227-1232.
2. Shahidi, F.; Finley, J. W. Omega-3 fatty acids: chemistry, nutrition and health effects. ACS Symposium Series 788. American Chemical Society, Washington, D.C., 2001.
3. Wanasundara, U.N.; Shahidi, F. *Food Chem.* **1998**, *63*, 335-342.
4. King, M. F.; Boyd, L. C.; Sheldon, B. W. *J. Am. Oil Chem. Soc.* **1992**, *69*, 545-551.
5. Reichert, R. D. *Trends Food Sci. Technol.* **2002**, *13*, 353-360.
6. Ferrari, R. A.; Schulte, E.; Esteves, W.; Brühl, L.; Mukherjee, K. D. *J. Am. Oil Chem. Soc.* **1996**, *73*, 587-592.
7. Lampi, A. M.; Hopia, A.; Piironen, V. Antioxidant activity of minor amounts of gamma-tocopherol on the oxidation of natural triacylglycerols, *J. Am. Oil Chem. Soc.* **1997**, *74*, 549-555.
8. Khan, M. A.; Shahidi, F. *J. Food Lipids* **1999**, *6*, 331-339.
9. Khan, M.A.; Shahidi, F. *J. Am. Oil Chem. Soc.* **2000**, *77*, 963-968.
10. Shahidi, F.; Lee, C. L.; Khan, M. A.; Barlow, P. J. Oxidative stability of edible oils as affected by their minor components. Presented at the Am. Oil Chem. Soc. Meeting & Expo, May 13-16, Minneapolis, MN, 2001.
11. Hamam, F.; Shahidi, F. *J. Am. Oil Chem. Soc.* Submitted for publication.
12. Hamam, F.; Shahidi, F. *J. Food Lipids* **2004**, In press.
13. Hamam, F.; Shahidi, F. *J. Agric. Food Chem.* **2004**.

Chapter 2

Formation of Odor-Active Carbonyls in Self-Assembly Structures of Phospholipids

I. Blank[1], J. Lin[2], M. E. Leser[1], and J. Löliger[1]

[1]Nestec Ltd., Nestlé Research Center, Vers-chez-les-Blanc, 1000 Lausanne 26, Switzerland
[2]Current address: Firmenich Inc., P.O. Box 5880, Princeton, NJ 08543

The formation of eight odor-active compounds generated from heated aqueous dispersions of phosphatidylcholine (PC) and phosphatidylethanolamine (PE) was studied. *trans*-4,5-Epoxy-(*E*)-2-decenal was found to be the most potent odorant on the basis of flavor dilution factors and odor activity values, followed by (*E,E*)-2,4-decadienal, 1-octen-3-one, and hexanal. The amount of (*E,E*)-2,4-decadienal in PC was about 20-fold higher compared to PE, while hexanal was the most abundant odor-active compound in the PE sample. Differences in the fatty acid composition of the phospholipids and the free amino group of PE can only partially explain the quantitative results found. It is suggested that the type of self-assembly structure adopted by phospholipid molecules in water is significantly influencing the reaction yields, thus playing a crucial role for the final quantitative composition of volatile constituents.

Introduction

Due to their amphiphilic nature, phospholipids are used as emulsifiers and stabilizers in food products, such as chocolate, baked products, shortenings, margarine, instant products, mayonnaise, and low-fat products (*1*). In general, oilseeds, cereal germs, egg yolk, and fish are the richest sources of phospholipids

(2, 3). Industrial phospholipids come almost entirely from soybeans with phosphatidylcholine (PC, lecithin), phosphatidylethanolamine (PE, cephalin), and phosphatidylinositol (PI) as major constituents of the phospholipid fraction (Figure 1). They are derivatives of phosphatidic acid (PA) and rich in polyunsaturated fatty acids (PUFAs), especially linoleic acid (C18:2), arachidonic acid (C20:4), and other highly unsaturated fatty acids (*e.g.* C22:5 and C22:6) (*4*).

These PUFAs are also a suitable source of odorants generated upon thermal treatment. Structural phospholipids have been shown to play a significant role in meat aroma specificity (*4*, *5*). The major volatile compounds found in heated phospholipids were hexanal, nonanal, 2-octenal, 2-decenal, (*E,E*)-2,4-decadienal, 1-octen-3-ol, 2-pentylfuran, and others (*6*). Significant differences were found between PC and PE on the basis of GC peak areas, in particular higher amounts of several unsaturated aldehydes in PC, while hexanal and 2-pentylfuran dominated in PE. The aroma composition of commercial soybean lecithin has been recently described, with lipid degradation and Maillard reaction products as the most potent odorants (*7*). (*E,E*)-2,4-Decadienal, (*E*)-2-nonenal, and 1-octen-3-one showed high sensory relevance (*8*).

In this contribution we report quantitative data of eight odor-active compounds generated in aqueous dispersions of phospholipids and discuss the role of reaction medium and structure with respect to the reaction yields.

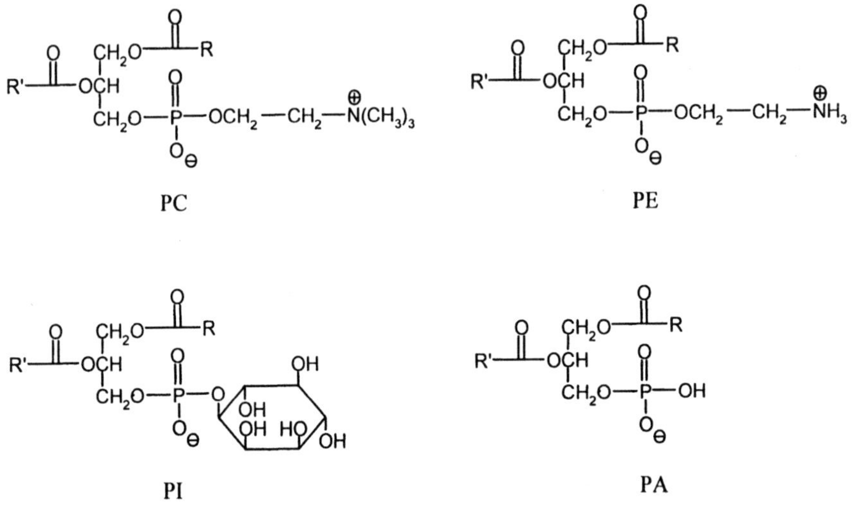

Figure 1. Chemical structures of phosphatidylcholine (PC, lecithin), phosphatidylethanolamine (PE, cephalin), phosphatidylinositol (PI), and phosphatidic acid (PA). The rest R represents the fatty acid chain (in the case of PA, the rest R was stearic acid).

Experimental

Materials

The following chemicals were commercially available: hexanal **1**, (*E*)-2-octenal **4**, (*E*,*E*)-2,4-nonadienal **9** (*E*,*Z*: 5%), 2-(1,1-dimethylethyl)-4-methoxyphenol (BHA), 2,6-di-*tert*-butyl-4-methylphenol (BHT), neutral aluminum oxide, methanol, (Aldrich/Fluka, Buchs, Switzerland); 1-octen-3-one **2** (Oxford, Brackley, UK); (*E*)-2-nonenal **7** (traces of *Z*-isomer **6**) (Agipal, Paris, France); (*E*,*E*)-2,4-decadienal **13** (*E*,*Z*: 5%, **12**), (*E*)-2-undecenal **14** (Fontarom, Cergy Pontoise, France); (*E*)-2-decenal **11** (Bedoukian, Danbury, Connecticut, USA); egg phosphatidylcholine (PC, >99%), egg phosphatidylethanolamine (PE, >99%) and distearoyl phosphatidic acid disodium salt (PA, >99%) (Avanti Polar Lipids, Copenhagen, Denmark); diethyl ether (Et_2O), hexane, pentane, silica gel 60, sodium chloride, anhydrous sodium sulfate (Merck, Darmstadt, Germany).

The following reference compounds and deuterated internal standards were synthesized: (*Z*)-1,5-octadien-3-one **3** (*9*), *trans*-4,5-epoxy-(*E*)-2-decenal **15** (*10*), (*E*,*Z*,*Z*)-2,4,7-tridecatrienal **16** (*11*), [5,6-$^2H_2$]-hexanal (***d-1***) (*12*), [1-$^2H_{1;2}$,2-$^2H_{1;1}$]-1-octen-3-one (***d-2***) (*9*), [2,3-$^2H_2$]-(*E*)-2-octenal (***d-4***) (*13*), [2,3-$^2H_2$]-(*E*)-2-nonenal (***d-7***) (*12*), [3,4-$^2H_2$]-(*E*,*E*)-2,4-nonadienal (***d-9***) (*12*), [3,4-$^2H_2$]-(*E*,*E*)-2,4-decadienal (***d-13***) (*12*), [4,5-$^2H_2$]-*trans*-4,5-epoxy-(*E*)-2-decenal (***d-15***) (*10*), [4,5,7,8-$^2H_4$]-(*E*,*Z*,*Z*)-2,4,7-tridecatrienal (***d-16***) (*13*).

Sample Preparation and Cleanup

Model Reactions. The solvent of a phospholipid chloroform-methanol solution (2:1, v/v, 10 mL) containing 1 g of egg PC or egg PE was evaporated with a stream of nitrogen. A phosphate buffer (50 mL, 0.5 M, pH 5.6) was then added and the mixture stirred magnetically to disperse the phospholipid. The sample was heated in a laboratory autoclave (Berghof, Eningen, Germany) for 30 min from room temperature to 145 °C, reaching the final temperature in 12 min with an average heating rate of 10 °C/min (Figure 2). After the reaction, the samples were rapidly cooled to room temperature with ice water.

Isolation of Volatile Compounds. For identification of odorants, the cooled reaction mixture was saturated with salt and the organic compounds continuously extracted with Et_2O (diethyl ether; 100 mL, containing 10 mg/L BHT (butylated hydroxytoluene) and BHA (butylated hydroxyanisole) as antioxidants) for 15 h using a liquid-liquid extractor. In the quantification experiments, defined amounts of labeled internal standards were added and mixed with the reaction mixtures before solvent extraction. Non-volatile compounds present in the solvent extract were removed by high vacuum transfer

(HVT) at 10^{-3} - 10^{-5} mbar (*14*). The condensates obtained by HVT were combined, dried over anhydrous sodium sulfate, and concentrated to 0.5 mL.

Column Chromatography (CC). Identification of odorants **6**, **15**, and **16** was achieved by fractionation at about 10 °C using a water-cooled glass column (20 x 1 cm) packed with a slurry of silica gel 60 (*14*). Elution was performed with 25 mL of each of the following pentane/Et$_2$O mixtures (v/v): 98/2 (fraction F1), 95/5 (fraction F2), 90/10 (fraction F3), 80/20 (fraction F4), and 50/50 (fraction F5). Each fraction was concentrated to 0.2 mL for analytical characterization. Odorants **6** and **16** were enriched in fraction F3, odorant **15** in F5.

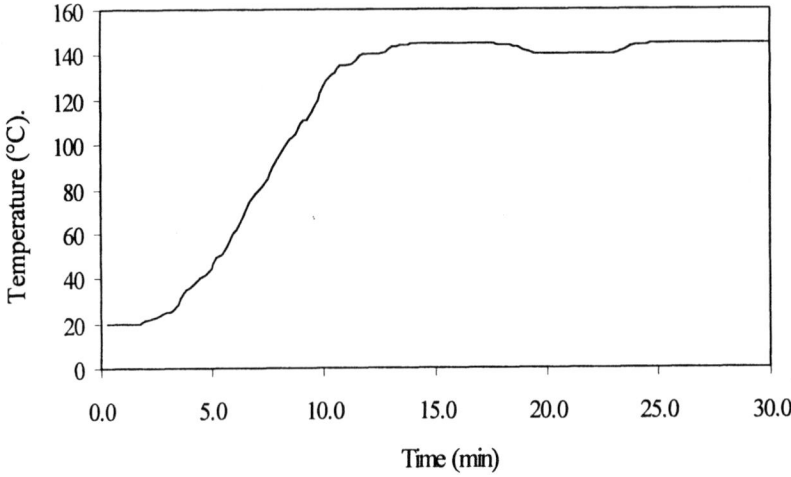

Figure 2. Temperature program for heating the model systems containing aqueous dispersions of phospholipids using a laboratory autoclave.

Chromatographic Techniques

Gas Chromatography-Olfactometry (GC-O). This was performed with a Carlo Erba Mega 2 gas chromatograph (Fisons Instruments, Schlieren, Switzerland) equipped with a cold on-column injector and a flame ionization detector (FID). Helium (80 kPa) was used as carrier gas. Fused silica capillary columns of low (DB-5) and medium (DB-1701) polarity were used, all 30 m x 0.32 mm with a film thickness of 0.25 µm (J&W Scientific, Folsom, CA). A splitter (Gerstel, Mülheim, Germany) was attached to the end of the capillary column to split the effluent 1:1 into the FID and sniffing port, both held at 230 °C, using deactivated and uncoated fused silica capillaries (50 cm x 0.32 mm).

The splitter was flushed with nitrogen (5 mL/min) to accelerate the gas flow. Just prior to the sniffing port, the GC effluent was mixed with humidified air (10 mL/min). Chromatographic conditions were used as described earlier (*12*). Linear retention indices (RI) were calculated according to van den Dool and Kratz (*15*).

Gas Chromatography - Mass Spectrometry (GC-MS). This was performed on a Finnigan MAT 8430 mass spectrometer (Bremen, Germany). Electron ionization (EI) mass spectra were generated at 70 eV. Chemical ionization (MS-CI) was performed at 150 eV with ammonia as the reagent gas. Further details of the GC-MS system and chromatographic conditions have been described previously (*12*). Relative abundances of the ions are given in percent.

Quantitative analysis was performed on a Finnigan SSQ 7000 mass spectrometer (Bremen, Germany) coupled with a HP-5890 gas chromatograph using isobutane as reagent gas for chemical ionization (CI) carried out at 200 eV. Further experimental details have been described previously (*13*). Quantitative measurements were carried out in full scan or in the selected ion monitoring mode. Each sample was prepared in duplicate and injected at least twice. The characteristic ions and GC-MS conditions used for quantification of **1, 2, 4, 7, 9, 13, 15**, and **16** by isotope dilution assays have been reported elsewhere (*13, 14*).

Results and Discussion

Odor-Active Compounds

The overall odor quality of heated aqueous dispersions of phospholipids was described as fatty, fried, metallic for PC and fishy, green, metallic for PE, which were very close to the overall odor of the corresponding aroma extracts. The GC chromatograms of heated PC and PE were similar and revealed several peaks present in both samples. As shown in Figure 3, the major volatile found in heated PC was compound **13**. The chromatogram of a control PC sample showed only traces of volatile compounds, thus indicating that most of them were generated upon heating.

GC-O was used as a screening method, resulting in sixteen odor-active regions, most of which were common to PC and PE (Figure 3). However, the sensory relevance of these odorants was different for the two samples as indicated by the FD-factors (Table 2). The fatty-fried smelling compound **13** was found to be the most potent odorant in the PC sample (FD= 500), followed by odorants **2, 7, 14, 15** and **16**. The aroma quality of all these odorants represented well the overall aroma of the PC sample (fatty, fried, metallic). On the other hand, odorant **2** was more pronounced in the PE sample, showing the highest FD factor (FD= 200). Further odor-active compounds with lower sensory relevance were the volatiles **13**, **15** and **16**. The sensory characteristics of these odorants were in good agreement with the overall aroma of the PE sample (fishy, metallic, green).

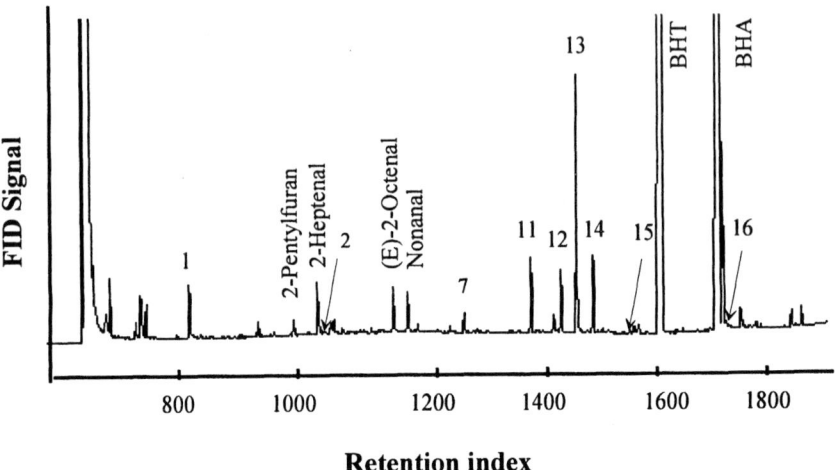

Figure 3. Volatile profile of a thermally treated aqueous dispersion of phosphatidylcholine.

Based on GC-O results, identification work was focused on odorants having high and medium FD-factors. The most intensely smelling odorant in PC was (E,E)-2,4-decadienal **13** (FD= 500), followed by (E)-2-undecenal **14** and *trans*-4,5-epoxy-(E)-2-decenal **15**. 1-Octen-3-one **2** (FD= 200) dominated in the PE sample followed by several odorants with lower FD-factors (Table I). Most of the odorants have been reported as volatile constituents of phospholipids, e.g. egg or soy lecithins (6, 7). However, (E)/(Z)-2-decenal **10/11**, (E)-2-undecenal **14**, and (E,Z,Z)-2,4,7-tridecatrienal **16** were identified for the first time in heated phospholipids.

Two odorants with FD= 50 that were more pronounced in the heated PE sample could not be identified due to the low concentration levels, i.e. odorant **5** (sweet) and odorant **8** (fishy). The major odorless volatile compounds generated in heated PC were (E)-2-heptenal, nonanal, and 2-pentylfuran (Figure 3), which did not smell at the concentration present in the sample.

Table I. Odorants Identified in the Aroma Extract of Heated Aqueous Dispersions of PC and PE

Odorant	Retention index		Odor quality	FD factor[d]	
	OV1701	SE-54	(GC-O)	PC	PE
Hexanal[a] 1	882	807	Green	1	1
1-Octen-3-one[a] 2	1069	986	Mushroom-like	50	200
(Z)-1,5-Octadien-3-one[b] 3	1087	979	Geranium-like	1	20
(E)-2-Octenal[a] 4	1173	1060	Fatty, soapy	1	1
(Z)-2-Nonenal[a,c] 6	1260	1151	Fatty, soapy	1	1
(E)-2-Nonenal[c] 7	1280	1164	Fatty, soapy	50	5
(E,E)-2,4-Nonadienal[a] 9	1355	1218	Fatty	5	5
(Z)-2-Decenal[a] 10	1365	1254	Soapy	10	1
(E)-2-Decenal[a] 11	1388	1267	Soapy	5	1
(E,Z)-2,4-Decadienal[a] 12	1436	1299	Fatty, soapy	20	5
(E,E)-2,4-Decadienal[a] 13	1461	1312	Fatty, fried	500	50
(E)-2-Undecenal[a] 14	1487	1369	Soapy	100	20
trans-4,5-Epoxy-(E)-2-decenal[a,c] 15	1564	1379	Metallic, green	100	100
(E,Z,Z)-2,4,7-Tridecatrienal[a,c] 16	1740	1589	Egg white-like	20	50

[a] Identification was based on reference compounds by GC-O and GC-MS. [b] Identification was based on reference compounds by GC-O only. The MS signals were too weak for unambiguous identification. [c] MS spectra were obtained after enrichment by column chromatography. [d] The original aroma extract having a flavor dilution (FD) factor of 1 was diluted stepwise with Et$_2$O until no odor-active region could be detected.

Formation of Selected Odorants from PC and PE

Quantitative Results. Eight compounds were selected for quantification according to their sensory relevance (FD factor) and chemical class to compare the potential of PC and PE in generating odorants when heated in aqueous dispersions (Figure 4), i.e. hexanal 1, 1-octen-3-one 2, (E)-2-octenal 4, (E)-2-noneal 7, (E,E)-2,4-nonadienal 9, (E,E)-2,4-decadienal 13, trans-4,5-epoxy-(E)-2-decenal 15, and (E,Z,Z)-2,4,7-tridecatrienal 16. As shown in Table II, concentrations varied from <1 mg/kg for 7 to about 100 mg/kg for 1 and 13. In the PC sample, 13 and 1 were the two most abundant odorants, followed by 4, 2, and 7. The amount of 1 generated in the PE sample was significantly higher (~100 mg/kg) than that of the other compounds, such as 4 and 13 for example (5-8 mg/kg).

In general, PC was more efficient in generating odorants, except of 1 that was preferably formed in PE. The most remarkable difference was found for the fatty smelling odorant **13**, the quantity of which was more than 20-fold higher in the PC sample. For the other odorants, the difference was less significant, *i.e.* about 2- to 5-fold. However, PE generated 1.5-fold more of the green smelling odorant **1** compared to PC.

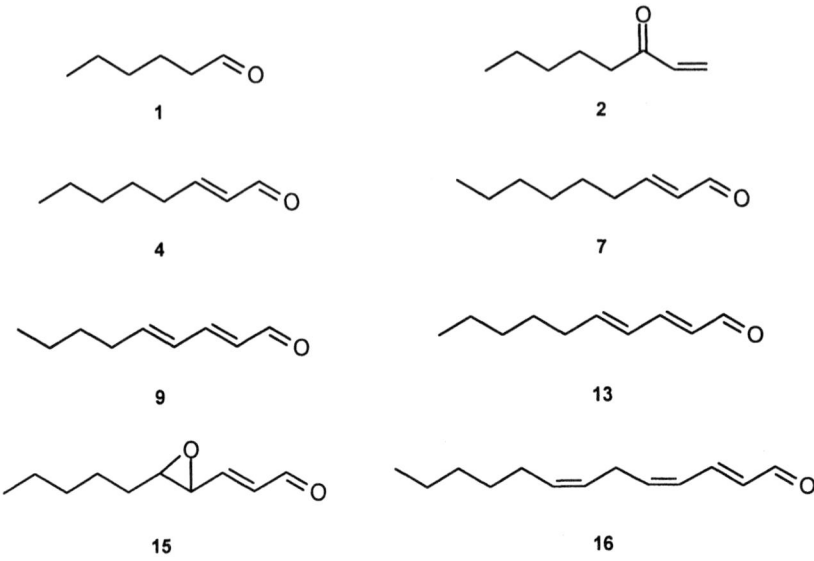

Figure 4. Chemical structure of the odorants quantified in aqueous dispersions of PC and PE.

Table II. Concentrations of Odorants Generated in Heated Aqueous Dispersions of PC, PE, and an Equimolar Mixture of PC and PE[a]

Odorant	PC	PE	PC + PE
1	60.7 ± 5.0	96.9 ± 7.4	88.9 ± 6.2
2	7.4 ± 2.4	1.8 ± 0.1	4.0 ± 0.1
4	20.2 ± 7.5	8.1 ± 0.2	4.2 ± 0.3
7	5.3 ± 1.4	2.5 ± 0.2	1.2 ± 0.1
9	1.2 ± 0.2	0.70 ± 0.04	0.41 ± 0.03
13	108.4 ± 28.4	4.8 ± 0.2	5.2 ± 0.8
15	4.6 ± 1.0	3.4 ± 0.7	1.0 ± 0.1
16	2.8 ± 0.5	1.7 ± 0.6	0.74 ± 0.14

[a] Concentration in mg/kg phospholipid.

Odor Activity Values (OAV). The sensory relevance of the phospholipid-derived carbonyls was estimated on the basis of OAV defined as the ratio of concentration to threshold value (*16*). As the concentrations of most of the volatile compounds identified in this study were above their odor thresholds, they likely contribute to the overall aroma of heated PC and PE. The OAV values were calculated by dividing the concentration of an odorant by its orthonasal threshold determined in oil (*14*).

As shown in Figure 5, odorant **15** (metallic) showed the highest OAV in both PC and PE. In PC, odorant **13** (fatty) and **2** (mushroom-like) were further important volatile compounds. In PE, odorant **1** (green) showed high OAV in addition to **2**, whereas the role of **13** was less pronounced. These odorants can be seen as the character impact compounds of heated phospholipids imparting metallic and fatty notes, which is in good agreement with the overall aroma of the heated samples. The fatty, fried, metallic note of heated PC is mainly imparted by odorants **13** and **15**, whereas odorants **1** and **2**, were additionally important for the fishy, green, metallic note of PE.

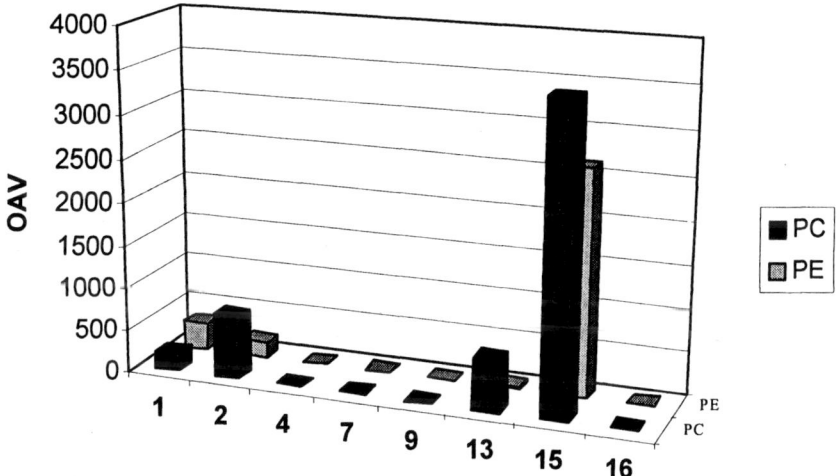

Figure 5. Odor Activity Values (OAV) of Odorants Generated in Heated Aqueous Dispersions of PC and PE.

The significant contribution of **15** is due to its extremely low odor threshold of 1.3 µg/L oil (*17*). Therefore, despite the low amounts found (Table 2), its sensory contribution is pronounced. On the contrary, the sensory relevance of **13** is mainly due to its high concentration, in particular in heated PC, in combination with a moderately low threshold value of 180 µg/L oil (*18*).

Precursors of Odorants. The compounds reported in this study are well-known lipid-degradation products of unsaturated fatty acids, such as oleic acid (C18:1), linoleic acid (C18:2), linolenic acid (C18:3), and arachidonic acid (C20:4) for example. The formation mechanisms have been described elsewhere *(19, 20).* Hexanal **1** can be generated from various fatty acids, such as C18:2 and C20:4. In addition, **1** is also known as a secondary autoxidation product of 2,4-decadienal **13** *(21, 22).* Linoleic and arachidonic acids are also direct precursors of odorant **13**. The formation of the newly identified (*E,Z,Z*)-2,4,7-tridecatrienal **16** in phospholipids can be explained by β-cleavage of the corresponding 8-hydroperoxy-5,9,11,14-eicosatetraenoic acid *(13).*

However, there is apparently no direct relationship between the amounts of **13** and **16** for example and the concentration of ω-6 fatty acids (*e.g.* C18:2 and C20:4) in the phospholipid, from which these aldehydes are formed. According to literature data *(4, 14),* the content of C20:4 in egg PE is several times higher than that in egg PC, while more **16** was generated in PC than in PE (Table III), thus indicating that the fatty acid composition is unlikely to be the only parameter explaining the formation of odorants from aqueous dispersions of phospholipids.

Table III. Fatty Acid Composition of Egg Phosphatidylcholine (PC) and Egg Phosphatidylethanolamine (PE)[a]

Fatty acid	Fatty acid composition (% total fatty acids)	
	egg PC	egg PE
C18:1 (n-9)	26.5	18.2
C18:2 (n-6)	16.7	13.2
C20:4 (n-6)	4.2	12.2
Σ Monounsaturated (1 DB)	29.3	17.7
Σ Diunsaturated (2 DB)	16.9	13.4
Σ Polyunsaturated (≥3 DB)	7.5	18.8

[a] Data were taken from the literature *(14).* DB= double bond.

Factors Affecting the Degradation of Phospholipids

Model Reactions Containing PC and PE. Quantitative characterization of an aqueous dispersion of equimolar amounts of PC and PE revealed hexanal **1** (~90 mg/kg) as the major odor-active constituent (Table II). The amounts of the remaining odorants varied from 0.4 mg/kg for (*E,E*)-2,4-nonadienal **9** to ~5 mg/kg for (*E,E*)-2,4-decadienal **13**. Surprisingly, the amounts of carbonyl odorants generated in the equimolar mixture of PC and PE (PC/PE 1:1) were very close to the levels generated from PE (Table 2). Compared to the quantitative data obtained with pure phospholipids, PC/PE 1:1 was almost identical with pure PE. The most intriguing difference, however, was the low amount of **13** found in PC/PE 1:1, while it was the major compound in pure PC.

Again, the fatty acid composition of PC and PE does not explain these findings. Assuming that odorant **13** is mainly formed from linoleic (C18:2) and arachidonic acid (C20:4), high amounts of **13** (~110 mg/kg) were found in the PC sample containing 20.9 % precursors. The amounts of **13** dropped to ~5 mg/kg in PE and PC/PE 1:1 containing of 25.4 % and 23.2 % precursors, respectively. These data confirm the hypothesis that several parameters affect the formation of carbonyls from heated aqueous dispersions of phospholipids.

Fatty Acid Composition. Surprisingly, the concentration of **16** is negatively correlated with the C20:4 content when comparing samples of pure PC and PE/PC mixtures (Figure 6). The highest amounts of **16** were found in pure PC (sample A), even though it contained the lowest level of C20:4 (4.2 %, Table III). Addition of PE to PC reduced the amounts of **16** (samples B-E). However, increasing PE in the PE/PC mixtures, and thus increasing the content of C20:4, led to more **16** generated. Finally, the C20:4 content of PE was about 3 times that of PC, however only about half the amount of **16** was formed (sample F).

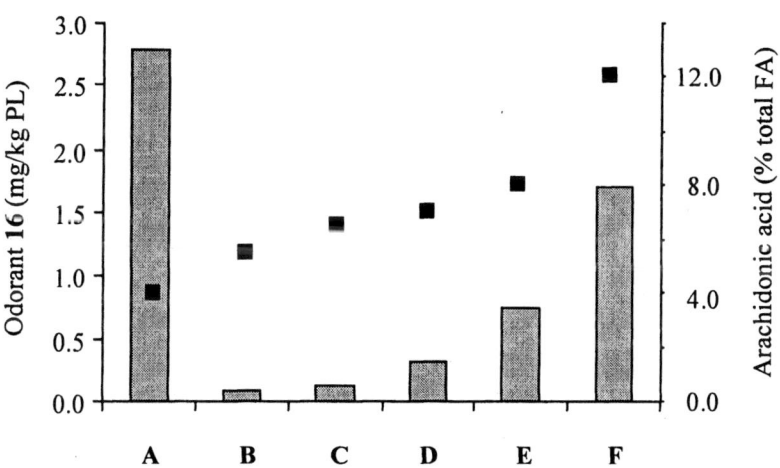

*Figure 6. Concentration of (E,Z,Z)-2,4,7-tridecatrienal **16** (bars, left vertical axis) generated upon heating of aqueous dispersions of phosphatidylcholine (PC), phosphatidylethanolamine (PE), and mixtures of PC and PE in various molar ratios along with the arachidonic acid content (■, right vertical axis) of the phospholipid samples. **A**, PC; **B**, PC+PE (4:1); **C**, PC+PE (2:1); **D**, PC+PE (3:2); **E**, PC+PE (1:1); **F**, PE.*

As shown in Figure 7, significantly (>20 fold) higher amounts of **13** were generated from pure PC (sample A) compared to the other phospholipid mixtures containing PE. Clearly, the presence of PE markedly reduced the concentration of **13**: already 25 % of PE in the phospholipid mixture (sample B) led to a decrease from ~90 mg/kg to less than 10 mg/kg. However, the total amounts of its direct precursors (C18:2, C20:4) increased from samples B to E with the rising level of PE, which is particularly rich in C20:4 (Table 3). The C18:2 content decreased slightly with increasing amounts of PE. This, however, does not explain the drastic decline in **13**. In the samples containing PE, the concentration of **13** decreased with decreasing C18:2 content (samples B and C) and increased slightly with increasing C20:4 content (samples C-F). These data suggest that there is no direct correlation between the amount of **13** and its precursors C18:2 and C20:4.

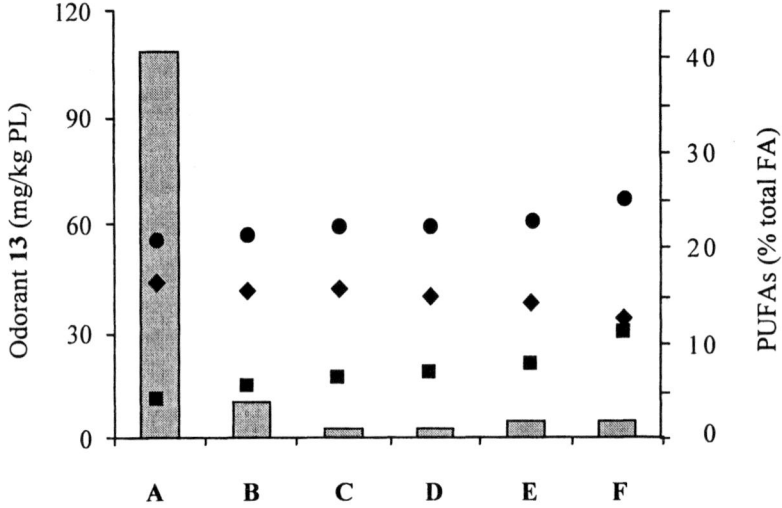

*Figure 7. Concentration of (E,E)-2,4-decadienal **13** (bars, left vertical axis) generated upon heating of aqueous dispersions of phosphatidylcholine (PC), phosphatidylethanolamine (PE), and mixtures of PC and PE in various molar ratios along with the content of linoleic acid (♦), arachidonic acid (■), and the sum of both (●) in the phospholipid samples (right vertical axis). **A**, PC; **B**, PC+PE (4:1); **C**, PC+PE (2:1); **D**, PC+PE (3:2); **E**, PC+PE (1:1); **F**, PE.*

The total unsaturated fatty acid contents of PC and PE are very close, *i.e.* about 52-54% (*14*). However, PE contains 2-3 times more PUFAs, known to readily oxidize. Surprisingly, the amounts of the odorants **13** and **16** generated from PE were lower than from PC, despite the higher PUFA content in PE.

However, as shown in Figure 6, a positive correlation between the concentration of **16** and C20:4 content was found in samples containing PE. This suggests, that already small quantities of PE change the reaction system leading to less unsaturated carbonyl odorants. Lower levels of unsaturated carbonyl compounds in PE-containing phospholipid samples were also reported in the literature (*6*). The authors speculated that reactions between aldehydes and the free amino group of PE were responsible for this phenomenon.

Polar Moiety. Indeed, the most remarkable difference between PC and PE is not the fatty acid composition, but the polar moiety, which can be involved in chemical reactions. For example, the primary amino group of PE may react with either the hydroperoxides or the aldehydes derived from them to give unstable Schiff bases, which may further react to give brown pigments (*23, 24*). As a result, such secondary reactions could contribute to degradation of odorants formed and, thus, affect the final levels of odorants.

The role of the polar moiety, in particular the free amino group of PE in reducing the amount of aldehydes by amino-carbonyl reactions, was studied by reacting PC and distearoyl phosphatidic acid (PA) in the molar ratio 4:1 and determining the concentration of odorant **13**. As shown in Table IV, not only PC/PE 4:1 affected the formation of **13** but also the presence of PA, *i.e.* the concentration of **13** dropped from ~110 mg/kg for PC to 10.1 and 25.4 mg/kg for the PC/PE (4:1) and PC/PA (4:1), respectively. These data suggest that the reactivity of the free amino group of PE does not sufficiently explain the drastic decrease in **13** by the chemical nature of the polar moiety of the phospholipid, nor the concentration of potential precursors present in the phospholipids.

Table IV. Concentration of (*E,E*)-2,4-Decadienal (13) Generated on Heating Aqueous Dispersions of PC and Mixtures of PC and PA or PE

Phospholipid (molar ratio)	*Concentration 13 (mg/kg)^a*	*Precursor content (% total fatty acids)*		
		C18:2^b	*C20:4^c*	*Σ*
PC	108.4	16.7	4.2	20.9
PC + PA (4:1)	25.4	13.4	3.4	16.8
PC + PE (4:1)	10.1	16.0	5.8	21.8

^{a} The concentration is given in mg/kg phospholipid. ^{b} C18:2, linoleic acid. ^{c} C20:4, arachidonic acid.

Molecular Organization. Phospholipids are amphiphilic molecules and act as emulsifier. They organize spontaneously forming a variety of different self-assembly structures in aqueous solutions, such as micelles, lamellar and reversed hexagonal phases (*25, 26*). When added to water, PC and PE induced significant differences in the appearance of the dispersions obtained, *i.e.* the reaction medium. While the PC sample was a homogeneous emulsion-like dispersion, the

PE sample became a solid block (lump), which did not disperse well into the water. This was the most striking difference between the PC sample and those containing PE, which may partially explain the differences found in the chemical composition of the lipid-derived odorants.

The difference in the appearance of the PC sample compared to those containing PE is the consequence of their binary phase behavior. According to the phase diagrams (Figure 8) (*27*), PC molecules adopt a lamellar phase, while a reversed hexagonal phase is formed in the PE sample under the experimental conditions used (2 % in water, 25-145 °C). The liquid crystalline lamellar phase is readily dispersible in water forming stable vesicles or liposomes with the so-called bilayer as basic self-assembly structure, in which the surfactant film has a zero curvature (*25*). On the contrary, PE molecules form a reverse hexagonal structure, in which the surfactant film is curved towards water. It is not possible to homogeneously disperse molecules organized in a reversed hexagonal phase (*28*), since this type of self-assembly structure is much more lipophilic, compared to the lamellar phase, due to the smaller polar moiety (area per headgroup) of the PE molecule. Pieces of such a phase fuse immediately when they come into contact with water, forming non-dispersible lumps.

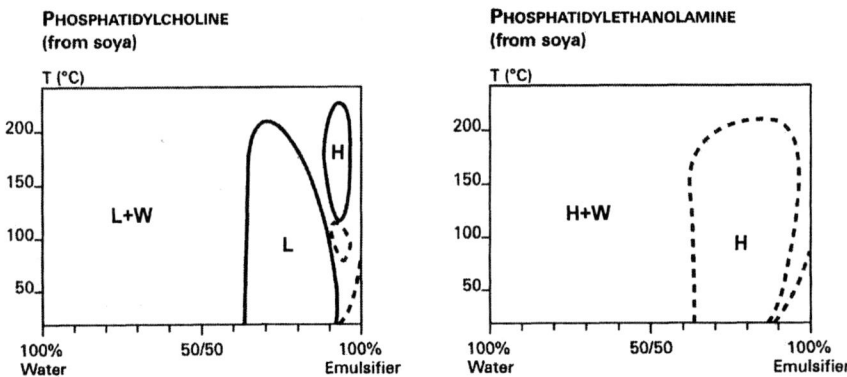

Figure 8. Phase diagrams of PC and PE. L, lamellar phase; L+W, lamellar phase and water; H, reversed hexagonal phase; H+W, reversed hexagonal phase and water. Adapted from ref. 28.

The ternary phase diagram of PC, PE and water obtained at room temperature is shown in Figure 9 (*25*), which should also be valid in this study using autoclave conditions (145°C), since the structure and phase behavior of aqueous dispersions containing pure phospholipid (PE or PC) is not affected by the temperature (Figure 8). The reversed hexagonal phase adopted by pure PE (Figure 6) is virtually dominating and unchanged until the addition of ~70 wt-%

of PC. Above 70 wt-%, the system separates into a three-phase area in which a reversed hexagonal phase is in equilibrium with a lamellar phase and water. This indicates that most probably in all PC/PE phospholipid mixtures used in this study (PC/PE 4:1, 2:1, 3:2 and 1:1) a reversed hexagonal phase was formed. These findings could explain the fact why both the formation of **13** and **16** was drastically reduced when adding PE to the PC. However, the mechanisms explaining why the formation of **13** and **16** is favored in a lamellar over a reversed hexagonal environment remains unclear.

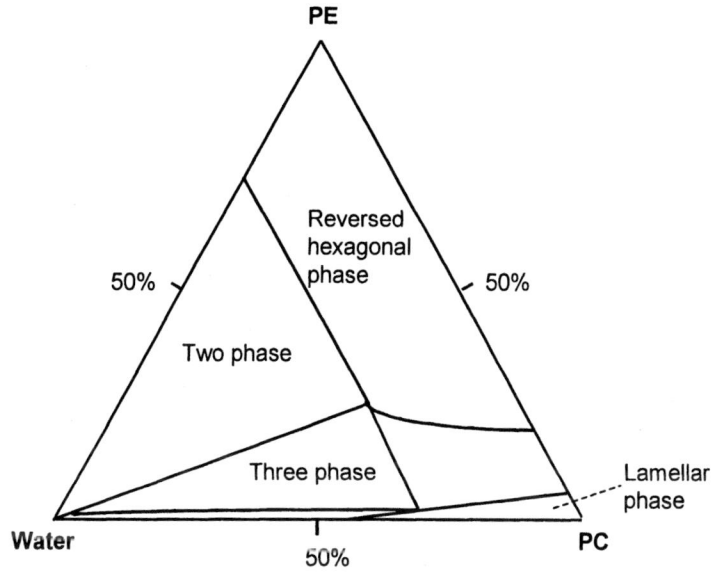

Figure 9. Ternary phase diagram of phosphatidylcholine (PC), phosphatidylethanolamine (PE), and water at 20 °C. Adapted from ref. 25.

In conclusion, our experimental findings support the hypothesis that the self-assembly structures adopted by the phospholipid molecules in aqueous media may be an important factor to consider in lipid oxidation. Additional work is required to elucidate the exact mechanisms explaining the differences observed in thermally-induced odorant formation in phospholipid self-assembly structures.

References

1. Gunstone, F. D. In *Structured and Modified Lipids*; Gunstone, F. D., Ed.; Marcel Dekker: New York, 2001; pp 241-250.
2. Cherry, J. P.; Kramer, W. H. In *Lecithins: Sources, Manufacture, and Uses*; Szuhaj, B. F.; Ed.; AOCS Press: Champaign, Il, 1989; pp 16-31.
3. Kuksis, A. In *Lecithins: Sources, Manufacture, and Uses*; Szuhaj, B. F., Ed.; AOCS Press: Champaign, IL, 1989; pp 32-71.
4. Farmer, L. J.; Mottram, D. S. *J. Sci. Food Agric.* **1990**, *53*, 505-525.
5. Mottram, D. S.; Edwards, R. A. *J. Sci. Food Agric.* **1983**, *34*, 517-522.
6. Farmer, L. J.; Mottram, D. S. *J. Sci. Food Agric.* **1992**, *60*, 489-497.
7. Stephan, A.; Steinhart, H. *J. Agric. Food Chem.* **1999**, *47*, 2854-2859.
8. Stephan, A.; Steinhart, H. *J. Agric. Food Chem.* **1999**, *47*, 4357-4364.
9. Lin, J.; Welti, H. D.; Arce Vera, F.; Fay, L. B.; Blank, I. *J. Agric. Food Chem.* **1999**, *47*, 2822-2829.
10. Lin, J.; Fay, L. B.; Welti, D. H.; Blank, I. *Lipids* **1999**, *34*, 1117-1126.
11. Blank, I.; Lin, J.; Arce Vera, F.; Welti, H. D.; Fay, L. B. *J. Agric. Food Chem.* **2001**, *49*, 2959-2965.
12. Lin, J.; Welti, D. H.; Arce Vera, F.; Fay, L. B.; Blank, I. *J. Agric. Food Chem.* **1999**, *47*, 2813-2821.
13. Lin, J.; Fay, L. B.; Welti, H. D.; Blank, I. *Lipids* **2001**, *36*, 749-756.
14. Lin, J.; Blank, I. *J. Agric. Food Chem.* **2003**, *51*, 4364-4369.
15. van den Dool, H.; Kratz, P. *J. Chromatogr.* **1963**, *11*, 463-471.
16. Grosch, W. *Flav. Fragr. J.* **1994**, *9*, 147-158.
17. Guth, H.; Grosch, W. *Lebensm. Wiss. Technol.* **1990**, *23*, 513-522.
18. Hall, G.; Anderson, J. *Lebensm. Wiss. Technol.* **1983**, *16*, 354-361.
19. Grosch, W. In *Autoxidation of Unsaturated Lipids*; Chan, H. W.-S., Ed.; Academic Press: London, 1987; pp. 95-139.
20. Frankel, E. N.; Neff, W. E.; Selke, E. *Lipids* **1981**, *16*, 279-285.
21. Matthews, R. F.; Scanlan, R.A.; Libbey, L. M. *J. Am. Oil Chem. Soc.* **1971**, *48*, 745-747.
22. Schieberle, P.; and Grosch, W. *J. Am. Oil Chem. Soc.* **1981**, *58*, 602-607.
23. Pokorny, J.; Tai, P. T.; Janicek, G. *Z. Lebensm. Unters. Forsch.* **1973**, *153*, 322-325.
24. Tai, P. T.; Pokorny, J.; Janicek, G. *Z. Lebensm. Unters. Forsch.* **1974**, *156*, 257-262.
25. Bergenstahl, B. A. In *Food Polymers, Gels and Colloids*; Dickinson, E., Ed.; Special Publication No. 82, The Royal Society of Chemistry at Norwich: Cambridge, 1991; pp 123-131.
26. Hawker, N. *Food Tech. Europe* **1994**, *1(5)*, 44-46.
27. Bergenstahl, B. A.; Claesson, P. M. In *Food Emulsions*; Marsson, K., Friberg, S. E., Eds.; Marcel Dekker: New York, NY, 1990; pp 41-96.
28. Larsson, K. In *Lipids - Molecular Organization, Physical Functions and Technical Applications*. The Oily Press Ltd.: Dundee, 1994; pp 107-113.

Chapter 3

The Effects of Diet, Breed, and Age of Animal at Slaughter on the Volatile Compounds of Grilled Beef

J. Stephen Elmore[1], Donald S. Mottram[1], Michael Enser[2], and Jeffrey D. Wood[2]

[1]School of Food Biosciences, The University of Reading, Whiteknights, Reading RG6 6AP, United Kingdom
[2]Division of Food and Animal Science, School of Veterinary Science, University of Bristol, Langford, Bristol BS40 5DU, United Kingdom

The aroma volatiles of grilled beef, from animals fed either grass silage or cereal concentrates, were compared. Aberdeen Angus and Holstein-Friesian cross-breed steers, slaughtered at 14 or 24 months, were studied. Compounds formed from linoleic acid, in particular 2-pentylfuran, 1-octen-3-ol, (Z)-2-octen-1-ol, and hexanal were at higher levels in the meat from the animals fed concentrates. Phytenes and compounds formed from α-linolenic acid, in particular 1-penten-3-ol and (Z)-2-penten-1-ol, were at higher levels in the meat of animals fed silage. Differences due to breed were small and not consistent with slaughter age. Dimethyl trisulfide, dimethyl disulfide and phenol were at higher levels in the meat of animals slaughtered at 24 months and may contribute to grilled beef aroma.

Ruminant diets based on forage are high in n–3 fatty acids, in particular α-linolenic acid (18:3 n–3), whereas concentrates-based diets are higher in n–6 fatty acids, in particular linoleic acid (18:2 n–6). The ratio of n–6 to n–3 fatty acids in concentrates-fed animals is 10, whereas in grass-fed animals it is only 1.3 (*1*).

The volatile composition of beef subcutaneous fat from animals fed forage is different to that from animals fed concentrates (*2*). Diterpenoids, derived from phytol, as well as 2,3-octanedione and 3-hydroxy-2-octanone, have been strongly associated with a forage diet, whereas lactones have been associated with a cereal diet (*3*). Saturated and monounsaturated aldehydes have been shown to be higher in cooked ground beef from cattle finished on concentrates (*4*). However, despite the differences that exist between the volatile compounds of beef from animals fed forage or cereal, Muir et al. (*5*) concluded that sensory panelists cannot reliably detect flavor differences between the two.

It is often assumed that the traditional breeds, such as Aberdeen Angus, have better flavor than dairy breeds, such as Holstein-Friesian. Vatansever et al. (*6*) compared meat from Welsh Black and Holstein-Friesian cross-breeds. The Welsh Black was significantly tougher than the Holstein-Friesian, but no differences in flavor were noted. Boylston et al. (*7*) showed that lipid-derived volatiles in Wagyu steers were higher than in three common American breeds, but only after storing the cooked meat at 3 °C for 3 days. The higher neutral lipids content of the Wagyu meat may have been responsible for this effect.

As part of continuing work to increase the n–3 content of ruminant meat, without compromising its flavor, this paper compares the volatile and fatty acid compositions of grilled beef from animals fed either concentrates or grass silage. The Aberdeen Angus, a typical beef breed, is compared with the Holstein-Friesian, a dairy breed, and the effect of slaughtering the animals at 14 or 24 months is examined. As far as we are aware, no work has been published on the effect of the animal's age at slaughter upon cooked beef aroma volatiles. A more comprehensive analysis of the animals slaughtered at 14 months is provided elsewhere (*8*).

Experimental

Meat Production and Cooking Conditions

Thirty-two Holstein-Friesian and 32 Aberdeen Angus steers were reared from 6 months of age until slaughter on either grass silage (*ad libitum* grass silage plus sugar beet pulp shreds at circa 15% of the total dry matter intake) or concentrates (a barley-based, full-fat soya concentrate and chopped barley straw at a ratio of 70:30 on a dry matter basis), resulting in 16 animals from each breed on each diet. The 18:2/18:3 ratio of the silage diet was 0.27 and that of the concentrates diet was 8.7. Animals were weighed every 14 days and this information was used to regulate the intake of the concentrates-fed animals, in order to maintain similar growth rates between diets within breed. Eight animals from each of the 4 treatment groups were slaughtered at 14 months of age and the rest at 24 months of age. After slaughter 2-cm samples of the M. *longissimus lumborum* were prepared, vacuum-packed and kept at −20 °C until analysis (3 months maximum).

Muscle samples from 6 animals from each of the 8 treatments (2 diets, 2 breeds, two ages at slaughter) were grilled and then their volatile compositions were analyzed by gas chromatography-mass spectrometry (GC-MS). Only 6 animals, randomly chosen from the 8 available, were analyzed, due to time constraints, and one analysis was performed for each animal. The frozen steaks were placed in a refrigerator at 4 °C, the day before they were to be cooked. Each steak was grilled separately under a pre-heated electric grill. Each sample, which comprised muscle surrounded by a layer of adipose tissue, was placed in the middle of the grilling tray in order to be grilled uniformly and cooked to a core temperature of 70 °C. Cooking to a fixed temperature compensated for any variations in the thickness of the samples, which ranged from approximately 2 cm to 3.5 cm. The steaks were turned over every two minutes. Directly after cooking, all adipose tissue was removed from the steak. The steak was then chopped in an electric bowl chopper. Forty grams of sample were taken for volatile analysis.

Analysis of Volatile Compounds

The minced steak was placed in a screw top conical flask (250 mL). A Dreschel head was attached to the flask, using an SVL fitting (Bibby, Stone, UK). The flask was held in a water bath at 60 °C for 1 h while nitrogen, at 40 ml/min, swept the volatiles onto a glass-lined, stainless steel trap (105 mm × 3 mm i.d.) containing 85 mg Tenax TA (Scientific Glass Engineering Ltd., Ringwood, Australia). A standard (100 ng 1,2-dichlorobenzene in 1 µl methanol) was added to the trap at the end of the collection and excess solvent and any water retained on the trap was removed by purging the trap with nitrogen at 40 ml/min for 10 min.

All analyses were performed on a Hewlett Packard 5972 mass spectrometer, fitted with a HP5890 Series II gas chromatograph and a G1034C Chemstation. A CHIS injection port (Scientific Glass Engineering Ltd.) was used to thermally desorb the volatiles from the Tenax trap onto a non-polar deactivated fused silica retention gap (5 m × 0.25 mm i.d.; Varian Chrompack International B.V., Middelburg, The Netherlands). The retention gap contained 5 small loops in a coil, which were cooled in solid carbon dioxide, contained within a 250-mL beaker. The retention gap was attached to a Supelcowax-10 fused silica capillary column (Carbowax 20M; 60 m × 0.25 mm i.d., 0.25 µm film thickness; Supelco, Bellefonte, PA). During the desorption period of 5 min, the oven was held at 40°C. After desorption, the oven was held at 40°C for a further 2 min before heating at 4°C/min to 250°C, where it was held for 10 min. Helium at 16 psi was used as the carrier gas, resulting in a flow of 1.0 ml/min at 40°C. A series of n-alkanes (C_5–C_{25}) in diethyl ether was analyzed, under the same conditions, to obtain linear retention index (LRI) values for the beef aroma components.

The mass spectrometer operated in electron impact mode with an electron energy of 70 eV and an emission current of 35 µA. The mass spectrometer scanned from m/z 29 to m/z 400 at 1.9 scans/s. Compounds were identified by first comparing their mass spectra with those contained in the NIST/EPA/NIH Mass Spectral Database or in previously published literature. Wherever possible identities were confirmed by comparison of linear retention index (LRI) values, with either those of authentic standards, or published values. Quantities of the volatile compounds were approximated by comparison of their peak areas with that of the 1,2-dichlorobenzene internal standard, obtained from the total ion chromatograms, using a response factor of 1.

Fatty Acid Analysis

The total fatty acid compositions of the loin muscles (*M. longissimus lumborum*) were measured. Fatty acids were analyzed by gas chromatography using the method of Choi *et al.* (*9*). Duplicate analyses were performed on six animals of each type.

Statistical Analysis

The starting date of the experiment for all of the treatments was the same. Hence, the animals slaughtered at 24 months were killed 10 months after the animals slaughtered at 14 months. Consequently, the animals slaughtered at 24 months were analyzed approximately 9 months later than the animals slaughtered at 14 months. Variability over time within the analytical equipment, such as column aging, wear within the injection port, ion source cleanliness, etc., means that a three-way analysis of variance may yield misleading results. Instead, two-way analysis of variance (ANOVA) was used separately on the two sets of data, to measure the effects of diet and breed on each compound identified in the GC-MS analyses of volatiles, and on the data for each muscle fatty acid. Data acquired for the animals slaughtered at 14 months was then compared with data acquired for the animals slaughtered at 24 months using the student t-test, in full knowledge of the limitations of this comparison, due to the variability in instrumental parameters described above.

Results and Discussion

Volatile Compounds

Fifty-one compounds were identified in the extract of at least one of the samples at or above a mean concentration of 10 ng per 100 g of grilled beef (Table I). These compounds were 13 hydrocarbons, 11 aldehydes, 9 alcohols, 7 ketones, 4 nitrogen-containing, 2 sulfur-containing, 2 furans, 1 ester and 2 ethers.

Two-way analysis of variance showed that 21 compounds were affected by diet and 4 compounds by breed in the animals slaughtered at 14 months (Table II). Of the compounds that were affected by diet, 17 were higher in the concentrates diet and 4 higher in the forage diet. In the animals slaughtered at 24 months, 15 compounds were affected by diet and 1 compound by breed (Table III). Of the compounds that were affected by diet, 10 were higher in the concentrates diet and 5 higher in the forage diet.

Of the compounds higher in the concentrates diet, 1-octen-3-ol, *cis*-2-octen-1-ol, 1-pentanol, 1-hexanol, pentanal, hexanal, heptanal, octanal and 2-pentylfuran have all been reported as oxidation products of C18:2 n–6 (*10*). Of the compounds higher in the forage diet 1-penten-3-ol, 2-penten-1-ol and 2-ethylfuran have been reported as oxidation products of C18:3 n–3 (*11*). 1-Phytene and 2-phytene, which are formed from the phytol moiety of chlorophyll, were also present at higher levels in the silage fed animals.

The only compound affected by breed at 24 months, 3-hydroxy-2-butanone was not affected by breed at 14 months. In fact, all breed effects were only significant at the 5% level and although 1-phytene and 2-phytene were present in the Holstein-Friesian at twice the amount compared with the Aberdeen Angus at 14 months, this effect was much smaller than the effect due to diet. 1-Phytene in particular was present in the forage-fed animals at 10 times the amount of that found in the concentrates-fed animals.

Larick *et al.* (*2-4*) showed that heated adipose tissue from steers fed forage contained higher levels of 1-phytene and 2-phytene and lower levels of heptane, pentanal, hexanal and 2-pentylfuran, in agreement with the data reported here. However, 2,3-octanedione, heptanal and 2-octene were reported by Larick's group as being lower in grain-fed steers than in grass-fed steers, which appears to contradict our results. However, the present work examined muscle tissue, which, with relatively high proportions of unsaturated lipid in the phospholipids, is likely to have a different total lipid profile to adipose tissue (*12*).

Table I. Compounds Identified in the Headspace of Grilled Beef

Hydrocarbons	*Aldehydes (cont.)*	*Nitrogen-containing (cont.)*
Hexane	Hexanal	Phenol
Heptane	Heptanal	
Octane	Octanal	*Furans*
2-Ethyl-1-hexene	Nonanal	2-Ethylfuran
(E)-2-Octene	Decanal	2-Pentylfuran
(Z)-2-Octene	Benzaldehyde	
Nonane		*Nitrogen-containing*
Benzene	*Ketones and*	Pyridine
2,2,4,6,6-Pentamethyl-heptane	*hydroxyketones*	Pyrazine
	Acetone	2,3-Dimethylpyrazine
Decane	2-Butanone	4,5-Dimethyl-2-pentyl-3-oxazoline
α-Pinene	2,3-Butanedione	
Toluene	2-Pentanone	Pyrrole
Ethylbenzene	2-Heptanone	Benzonitrile
1,3-Dimethylbenzene	3-Hydroxy-2-butanone	2,4,5-Trimethyl-imidazole
1,2-Dimethylbenzene	2,3-Octanedione	
Limonene	6-Methyl-5-hepten-2-one	4,5-Dimethyl-2-isobutylimidazole
Styrene		
Tetradecane	Acetophenone	4,5-Dimethyl-2-butylimidazole
Pentadecane		
Hexadecane	*Alcohols*	4,5-Dimethyl-2-pentylimidazole
Heptadecane	2-Methyl-1-propanol	
1-Phytene	1-Butanol	
Octadecane	1-Penten-3-ol	*Sulfur-containing*
2-Phytene	1-Pentanol	Hydrogen sulfide
	3-Methyl-3-buten-1-ol	S-Methyl thioacetate
	(Z)-2-Penten-1-ol	Dimethyl disulfide
Aldehydes	3-Methyl-2-buten-1-ol	Dimethyl trisulfide
2-Methylpropanal	1-Hexanol	
Butanal	1-Octen-3-ol	*Miscellaneous*
2-Methylbutanal	1-Heptanol	Ethyl acetate
3-Methylbutanal	1-Octanol	(1,1-Dimethylethoxy)-benzene
Pentanal	(Z)-2-Octen-1-ol	

Table II. Aroma Compounds of Grilled Steaks from Steers Slaughtered at 14 Months of Age, which were Significantly Affected by Diet or Breed

Compound [a]	concentration in headspace (ng/100g) [b]				P_{DIET} [c]	P_{BREED} [d]	LRI [e]
	Aberdeen Angus		Holstein-Friesian				
	silage	conc.	silage	conc.			
Heptane	43	59	25	73	*	NS	700
(E)-2-Octene	11	34	8	33	***	NS	854
(Z)-2-Octene	5	16	4	16	***	NS	867
Pentanal	180	270	110	309	**	NS	987
S-Methyl thioacetate	25	17	42	25	*	*	1062
Hexanal	360	1300	300	1600	***	NS	1096
1-Penten-3-ol	82	45	66	54	*	NS	1174
Heptanal	120	190	84	240	**	NS	1197
2-Pentylfuran	17	33	6	42	**	NS	1236
1-Pentanol	100	180	75	190	**	NS	1255
Octanal	86	110	69	150	*	NS	1293
(Z)-2-Penten-1-ol	7	4	6	5	*	NS	1321
2,3-Octanedione	8	13	10	24	*	NS	1328
1-Hexanol	18	30	14	39	**	NS	1351
1-Octen-3-ol	32	97	35	150	***	NS	1447
4,5-Dimethyl-2-pentyl-3-oxazoline	1	7	2	11	**	NS	1514
(Z)-2-Octen-1-ol	2	4	2	8	**	NS	1614
Benzonitrile	2	1	1	1	*	NS	1626
Acetophenone	1	1	1	1	NS	*	1671
1-Phytene	22	2	43	4	***	*	1792
2-Phytene	3	1	6	4	*	*	1871
4,5-Dimethyl-2-isobutylimidazole	6	3	6	2	*	NS	2266

[a] The mass spectrum and LRI of compounds in italics agree with those of the authentic compound.

[b] Means are from six samples from six different animals.

[c, d] Probability that there is a difference between means for (c) diet and (d) breed; NS, no difference between means ($P > 0.05$); * significant at the 5% level; ** significant at the 1% level; *** significant at the 0.1% level.

[e] Linear retention index on a Supelcowax-10 column.

Table III. Aroma Compounds of Grilled Steaks from Steers Slaughtered at 24 Months of Age, which were Significantly Affected by Diet or Breed

Compound [a]	concentration in headspace (ng/100g) [b]				P_{DIET} [c]	P_{BREED} [d]	LRI [e]
	Aberdeen Angus		Holstein-Friesian				
	silage	conc.	silage	conc.			
(E)-2-Octene	4	8	6	7	*	NS	854
(Z)-2-Octene	2	4	3	4	*	NS	867
2-Ethylfuran	6	5	7	3	*	NS	956
Pentanal	191	369	182	240	*	NS	987
Hexanal	214	1983	386	1429	***	NS	1096
1-Penten-3-ol	81	47	113	39	**	NS	1174
2-Pentylfuran	5	24	7	21	***	NS	1236
3-Hydroxy-2-butanone	2917	2253	4625	3418	NS	*	1300
(Z)-2-Penten-1-ol	10	7	16	5	**	NS	1321
2,3-Octanedione	6	21	9	18	**	NS	1328
1-Hexanol	47	81	38	71	*	NS	1351
1-Octen-3-ol	33	237	55	203	***	NS	1447
(Z)-2-Octen-1-ol	3	16	6	14	***	NS	1614
1-Phytene	313	14	364	64	***	NS	1792
2-Phytene	49	19	54	39	**	NS	1871
4,5-Dimethyl-2-pentylimidazole	25	44	17	42	*	NS	2491

Footnotes: see Table II.

The compounds that showed the greatest differences between treatments are likely to have the biggest effect on flavor differences between the steaks. At 14 months 1-octen-3-ol, (Z)-2-octen-1-ol, hexanal, 2-pentylfuran, *cis*- and *trans*-2-octene and 4,5-dimethyl-2-pentyl-3-oxazoline were all over 3 times higher in the concentrates-fed steers, compared with the grass-fed steers. At 24 months 1-octen-3-ol, (Z)-2-octen-1-ol, hexanal and 2-pentylfuran were all over 3 times higher in the concentrates-fed steers, compared with the grass-fed steers. 1-Phytene was present at much higher levels in grass-fed beef compared with concentrates-fed beef at both 14 and 24 months.

Several compounds were present at substantially higher levels in the 24-month beef compared with the 14-month beef, even allowing for variation in instrumental conditions with time (Table IV). Dimethyl disulfide and dimethyl trisulfide were present at levels over ten times higher in the 24-month beef, compared with the 14-month beef, as were phenol and a compound tentatively identified as 2,4,5-trimethylimidazole, which has not been previously reported in food. Dimethyl trisulfide is an important contributor to beef aroma. Aroma properties attributed to this compound in cooked beef include musty, roasty, rubbery, cabbage-like, burnt and gravy-like *(13-15)*. Dimethyl disulfide contributed moldy, pungent, rubbery, onion-like notes to shallow-fried beef *(14)*. 4,5-Dimethyl-2-pentyl-3-oxazoline is the only compound that was found at much higher levels in the 14 month steaks. This compound has not been previously reported in food.

Fatty acids

As expected, all of the fatty acids in the muscle were affected by dietary treatment (Tables V and VI). At both slaughter ages the ratio of polyunsaturated fatty acids to saturated fatty acids (the P:S ratio) and the $n-6$:$n-3$ ratios were higher in the concentrates-fed animals, whereas total lipids were higher in the silage-fed animals. At 14 months all of the saturated and monounsaturated fatty acids were significantly affected by diet at the 0.1% level, whereas at 24 months, although differences were still significant, most were at the 5% level, with only 16:1 significant at the 1% level.

With regard to breed, the P:S ratio was higher in the Holstein-Friesian cattle compared to the Aberdeen Angus cattle at both slaughter ages, but no other effects of breed were observed at 14 months. However, at 24 months slaughter age the saturated acids palmitic (16:0) and stearic (18:0) were higher in the Angus steers, whereas three of the longer chain $n-3$ fatty acids were present at higher levels in the Holstein-Friesians. Even so, most differences in fatty acid composition were due to diet.

Table IV. Effect of Slaughter Age on Selected Aroma Compounds of Grilled Beef Steaks

Compound[a]	concentration in headspace (ng/100g)[b]								LRI[c]
	14 months				24 months				
	Aberdeen Angus		Holstein-Friesian		Aberdeen Angus		Holstein-Friesian		
	silage	conc.	silage	conc.	silage	conc.	silage	conc.	
Dimethyl disulfide	5	2	2	7	73	64	45	39	1090
Dimethyl trisulfide	3	2	1	7	79	44	50	25	1394
4,5-Dimethyl-2-pentyl-3-oxazoline	1	7	2	11	1	–	–	–	1514
Acetophenone	1	1	1	1	15	12	9	11	1671
1-Phytene	22	2	43	4	313	14	364	64	1792
Octadecane	1	1	4	2	19	17	25	19	1800
2-Phytene	3	1	6	4	49	19	54	39	1871
Phenol	1	1	2	2	26	24	16	20	2016
2,4,5-Trimethylimidazole	3	3	13	7	116	65	87	45	2237
4,5-Dimethyl-2-isobutylimidazole	6	3	6	2	41	18	17	19	2266
4,5-Dimethyl-2-butylimidazole	2	4	3	2	32	30	14	21	2383
4,5-Dimethyl-2-pentylimidazole	3	10	4	4	25	44	17	42	2491

[a] The mass spectrum and LRI of compounds in italics agree with those of the authentic compound.
[b] Means are from six replicate samples from six different animals.
[c] Linear retention index on a Supelcowax-10 column.

Table V. Total Lipid Composition of Muscle (*longissimus lumborum*) of Aberdeen Angus and Holstein-Friesian Steers Fed Diets Containing Silage or Concentrates (conc.) and Slaughtered at 14 Months of Age

fatty acid	fatty acids (mg/100g muscle)[a]				P_{DIET}[b]	P_{BREED}[c]
	Aberdeen Angus		Holstein-Friesian			
	silage	conc.	silage	conc.		
14:0	63	35	61	30	***	NS
16:0	687	426	691	348	***	NS
16:1	103	48	88	40	***	NS
18:0	334	255	356	222	***	NS
18:1 *n*–9	906	531	904	453	***	NS
18:1 *n*–7	25	47	29	44	***	NS
18:2 *n*–6	62	140	66	146	***	NS
18:3 *n*–3	30	7.2	36	6.7	***	NS
20:3 *n*–6	6.1	15	6.4	14	***	NS
20:4 *n*–6	33	54	36	51	***	NS
20:4 *n*–3	4.3	0.82	7.5	0.72	**	NS
20:5 *n*–3	18	4.7	19	4.0	***	NS
22:4 *n*–6	2.1	6.3	2.4	6.4	***	NS
22:5 *n*–3	24	11	26	9.7	***	NS
22:6 *n*–3	4.4	1.3	6.1	1.2	***	NS
TOTAL	2600	1800	2680	1570	***	NS
P:S[d]	0.087	0.21	0.097	0.26	***	*
Σ*n*–6:Σ*n*–3[e]	1.3	8.5	1.2	9.3	***	NS

[a] Means are from six replicate samples from six animals.

[b,c] Probability that there is a difference between means for (*b*) diet and (*c*) breed; NS, no significant difference between means ($P > 0.05$); * significant at the 5% level; ** significant at the 1% level; *** significant at the 0.1% level.

[d] P:S ratio is (18:2 *n*–6 + 18:3 *n*–3)/(14:0 + 16:0 + 18:0).

[e] Σ*n*–6 consists of 18:2, 20:3, 20:4; Σ*n*–3 consists of 18:3, 20:4, 20:5, 22:5, 22:6.

Table VI. Total Lipid Composition of Muscle (*longissimus lumborum*) of Aberdeen Angus and Holstein-Friesian Steers Fed Diets Containing Silage or Concentrates (Conc.) and Slaughtered at 24 Months of Age

fatty acid	fatty acids (mg/100g muscle)[a]				P_{DIET}[b]	P_{BREED}[c]
	Aberdeen Angus		Holstein-Friesian			
	silage	conc.	silage	conc.		
14:0	360	131	150	117	*	NS
16:0	3153	1346	1488	1239	*	*
16:1	505	212	274	204	**	NS
18:0	1266	592	639	607	*	*
18:1 n–9	4015	1805	2159	1754	*	NS
18:1 n–7	187	104	118	95	*	NS
18:2 n–6	91	193	93	216	***	NS
18:3 n–3	60	7.9	54	17	***	NS
20:3 n–6	5.3	18	8.1	19	***	NS
20:4 n–6	25	69	38	78	***	NS
20:4 n–3	5.7	1.1	7.2	2.0	***	NS
20:5 n–3	18	1.7	33	7.2	***	**
22:4 n–6	1.6	12	1.9	8	***	NS
22:5 n–3	26	7.4	36	12	***	*
22:6 n–3	3.4	0.5	7.7	1.1	***	**
Total	9743	4527	5119	4399	*	*
P:S[d]	0.032	0.097	0.065	0.119	***	*
Σn–6:Σn–3[e]	1.1	15.7	1.0	8.2	***	NS

Footnotes: see Table V.

Inevitably, the concentrations of all muscle fatty acids increased from 14 to 24 months. However, the biggest increases were in the more saturated acids, which increased by up to four times between 14 and 24 months, whereas concentrations of 18:2 n–6 and 18:3 n–3 doubled at most. Concentrations of some of the long-chain polyunsaturated fatty acids (longer than 20 carbon atoms) hardly increased at all. This is reflected in the average P:S values at 14 and 24 months, which are 0.16 and 0.078 respectively.

Conclusions

Volatile compounds present in headspace extracts from grilled beef can reveal whether the animal was fed a diet high in n–6 or n–3 fatty acids. The effect of breed on beef flavor appears to be limited but there appear to be interesting aroma compounds developing in the beef of cattle as they age. Another experiment, where all of the feeding trials finish at the same time, would reveal to a greater degree how the age of the animal at slaughter affects the aroma composition of its meat.

Acknowledgements

We are most grateful to K. G. Hallett for fatty acid determinations and to N. D. Scollan for conducting the feeding trials. This work was funded by a LINK project (Project No. LK0644) involving the Department of Environment, Food and Rural Affairs, the Meat and Livestock Commission, and a consortium of companies.

References

1. Enser, M.; Hallett, K. G.; Hewett, B.; Fursey, G. A. J.; Wood, J. D.; Harrington, G. *Meat Sci.* **1998**, *49*, 329-341.
2. Larick, D. K.; Hedrick, H. B.; Bailey, M. E.; Williams, J. E.; Hancock, D. L.; Garner, G. B.; Morrow, R. E. *J. Food Sci.* **1987**, *52*, 245-251.
3. Maruri, J. L.; & Larick, D. K. *J. Food Sci.* **1992**, *57*, 1275-1281.
4. Larick, D. K.; Turner, B. E. *J. Food Sci.* **1990**, *54*, 649-654.
5. Muir, P. D.; Deaker, J. M.; Bown, M. D. *N. Z. J. Agric. Res.* **1998**, *41*, 623-635.
6. Vatansever, L.; Kurt, E.; Enser, M.; Nute, G. R.; Scollan, N. D.; Wood, J. D.; Richardson, R.I. *Anim. Sci.* **2000**, *71*, 471-482.
7. Boylston, T. D.; Morgan, S. A.; Johnsen, K. A.; Wright Jr.; R. W.; Busboom, J. R.; Reeves, J. J. *J. Agric. Food Chem.* **1996**, *44*, 1091-1095.

8. Elmore, J. S.; Warren, H. E.; Mottram, D. S.; Scollan, N. D.; Enser, M.; Richardson, R. I.; Wood, J. D. *Meat Sci.* **2004**, *68*, 27-33.
9. Choi, N. J.; Enser, M.; Wood, J. D.; Scollan, N. D. *Anim. Sci.* **2000**, *71*, 509-519.
10. Grosch, W. In *Autoxidation of Unsaturated Lipids*; Chan, H.W.-S., Ed.; Academic Press: London, 1987; pp 95-139
11. Elmore, J.S.; Campo, M.M.; Enser, M.; Mottram, D.S. *J. Agric. Food Chem.* **2002**, *50*, 1126-1132.
12. Enser, M.; Hallett, K.G.; Hewett, B.; Fursey, G.A.J.; Wood, J.D. *Meat Sci.* **1996**, *42*, 443-456.
13. Gasser, U.; Grosch, W. *Z. Lebensm.-Unters. -Forsch.* **1988**, *186*, 489-494.
14. Specht, K.; Baltes, W. *J. Agric. Food Chem.* **1994**, *42*, 2246-2253.
15. Machiels, D.; van Ruth, S. M.; Posthumus, M. A.; Istasse, L. *Talanta* **2003**, *60*, 755-764.

Chapter 4

Flavor Release from French Fries

Wil A. M. van Loon[1], Jozef P. H. Linssen[1],
Alexandra E. M. Boelrijk[2], Maurits J. M. Burgering[2],
and Alphons G. J. Voragen[1]

[1]Laboratory of Food Chemistry, Department of Agrotechnology and Food Sciences, Wageningen University, Wageningen, The Netherlands
[2]NIZO Food Research, Ede, The Netherlands

Flavor release from French fries was measured with the MS-NOSE using both panelists and a mouth-model system. The identity of several volatiles measured with the MS-NOSE was verified with MS-MS. The effect of frying time and the effect of adding salt on I_{max} (maximum intensity of compounds) and on t_{max} (time of maximum intensity of compounds) were determined. I_{max} of the formation of all compounds correlated with frying time. Addition of salt resulted in a lower t_{max}, but no significant effect on I_{max} was found. *In vivo* measurements with panelists showed that all components reached t_{max} within 10 seconds, while *in vitro* measurements with the mouth model system showed that low molecular compounds reached t_{max} within 50 seconds, while higher molecular compounds reached t_{max} after 3-5 minutes.

Introduction

French fries are appreciated throughout the world for their pleasant taste and texture. The flavor of potato and potato products has been investigated extensively and more than 500 volatiles have been identified (*1, 2*). Most of the research focused on volatile compounds of raw, boiled, (microwave) baked potato, and potato chips (*3-14*). Only a few papers have been published about volatile compounds of French fries (*15, 16*).

The compositions of the volatiles obtained from the different processes differ significantly (2). Heating changes the profile of volatiles of boiled potato in comparison with raw potato (1). At temperatures above 100°C heterocyclic compounds (e.g. pyrazines) are formed (2) and use of oil introduces new compounds formed by interaction of lipids and Maillard-reaction products (17).

For baked potatoes, potato chips and French fries, on a quantitative basis, most odor compounds derive from lipid degradation and Maillard reaction and/or sugar degradation (8). Wagner and Grosch (16, 18) identified 48 odorants of French fries by application of both Aroma Extraction Dilution Analysis (AEDA) and GC-Olfactometry.

With the introduction of MS-NOSE it became possible to measure volatiles in the nose during eating (19). Most of the studies using MS-NOSE focused on model systems (20-22). Flavor release would give information about the sensory perception during eating French fries, but this has not been studied until now. Therefore, the aim of this study was to follow the release of flavor from French fries in real time.

Materials and Methods

Materials

Partially fried frozen French fries (cv. Agria, 10x10mm) were kindly provided by Boots Frites BV (Purmerend, The Netherlands). Partially hydrogenated vegetable oil (Remia, Den Dolder, The Netherlands) used for frying was obtained from a local supplier. French fries of about 6 cm were selected and finish-fried individually in a household fryer (Princess 2620, Breda, The Netherlands) at 180°C (356°F).

Artificial saliva was prepared in demineralized water according to Van Ruth et al. (23) and consisted of K_2PO_4 (1.37 g/L), KCl (0.45 g/L), $CaCl_2 \cdot H_2O$ (0.44 g/L), NaCl (0.88 g/L), $NaHCO_3$ (5.2 g/L), porcine stomach mucine (2.16 g/L, type II M2378, Sigma, Steinheim, Germany) and α-amylase from A. oryzae (10.5 g/L, type X-A, 500000 units, Sigma, Steinheim, Germany). NaN_3 (2 ml of 10% solution) was used for preservation.

General Setup

After finish-frying flavor release from French fries was measured by MS-NOSE both in exhaled breath of panelists (*in vivo*) and in a mouth model system (*in vitro*). The effect of frying time (2, 4, 6, and 8 min) and addition of salt (0.1

g) on maximum intensity (I_{max}) and time of maximum intensity (t_{max}) of flavor release were evaluated.

Identification of Released Flavors

Concentrations of flavor compounds were measured on-line by an atmospheric pressure chemical ionization (APCI) gas phase analyzer attached to a VG Quattro II mass spectrometer (Micromass UK Ltd., Manchester, UK). Compounds were ionized by a 3.0 kV discharge (source and probe temperatures were 80°C), and scanned for m/z 40 – 250. M/z-values of observed compounds were selected and fragmented with argon for identification.

***In vivo* Flavor Release in Exhaled Breath**

Flavor release was measured in exhaled breath of three experienced panelists in triplicate. Panelists breathed in and out slowly through a tube in the nose, from which continuously 80 mL/min of air was sampled directly into the APCI-MS. A strict protocol was followed during experiments. After putting one French fry in the mouth, panelists immediately started chewing at about one chewing movement per second. The sample was swallowed after 30 s and chewing was continued until 60 seconds. Between French fries the mouth was rinsed with water. Blank experiments were recorded with water following the same protocol. Acetone, present in human breath from fatty acid metabolism, was measured as indicator for the breathing pattern.

***In vitro* Flavor Release in the Mouth Model System**

Dynamic headspace measurements were carried out in triplicate with a mouth model system developed by Van Ruth (*24*). The mouth model consists of a double wall glass housing with an inner volume comparable to the human mouth in which a plunger moves up and down and rotates simultaneously. Water of 37°C (99°F) is pumped through the double wall.

One French fry was put in the mouth model system and 3.5 mL of artificial saliva was added. The amount of artificial saliva was determined from the weight difference before and after chewing a French fry for 30 s. The mouth model system "chewed" at 1.6 Hz and 80 mL/min of air was sampled directly into the APCI-MS. Flavor release was monitored for 5 min immediately after starting "chewing".

Statistical Analysis

SPSS 10.0.7 was used for statistical evaluation of the data. Linear regression was used for the effect of frying time, and MANOVA for the effect of salt addition. Differences were regarded significant if $\alpha < 0.05$ and $\beta > 0.20$.

Results and Discussion

Identification of observed compounds

Release of a total of 11 compounds could be observed with the mouth model system, 5 of these were found also *in vivo* (table 1). More compounds were found *in vitro*, because the whole volume released in the mouth model system entered the APCI-MS, while only a small part of the breath was sampled from the panelists. For a number of compounds this small part was not high enough to reach the detection limit.

In accordance with literature (*8*), the observed compounds originate from either lipid degradation or Maillard reaction. Methylpropanal, 3-methylbutanal, and 2-methylbutanal are Strecker aldehydes from valine, leucine, and isoleucine respectively (*17*) and pyrazines are known Maillard reaction products as well (*25, 26*). 2-heptenal and 2,4-decadienal are formed from autoxidation of linoleic acid (*27, 28*). The fragmentations of m/z 75 and m/z 91 were similar to 1-hydroxy-2-propanone and glyceraldehyde, respectively. Glyceraldehyde has not been described in literature as a volatile compound from potato or potato products previously, and 1-hydroxy-2-propanone was described only once (*1*). Both are however Maillard reaction intermediates (*29*). M/z-value 69 could not be identified. Its fragmentation showed similarities with furan and pyrazole, but both have not been reported in potato or potato products previously. Because the fragmentation pattern of m/z 87 showed a high peak of m/z 69, it is also possible that m/z 69 is a fragment of 2- or 3-methylbutanal.

Due to low quantities it is likely that we were not able to detect trace compounds with low threshold values, which play a role in odor or taste sensation.

Effect of frying time on I_{max} and t_{max}

For all compounds I_{max} increased linearly with frying time (figure 1). Maximum intensity was higher *in vitro* than *in vivo*, because all compounds released in the mouth model system entered the APCI-MS, while only a small part was sampled from the breath of panelists. R^2 of linear regression was higher *in vitro* (0.60-0.85) than *in vivo* (0.25-0.55). This can be explained by differences among panelists such as geometry of the mouth, saliva production, and chewing behavior. Highest I_{max} values were reached for methylpropanal, 2- and 3-methylbutanal. Martin and Ames (*25*) reported high concentrations of 2- and 3-methylbutanal in fried potato model systems compared to pyrazines. Methylpropanal was not mentioned, probably because valine was not included in the model systems.

No significant effect of frying time on t_{max} could be found (figure 2). For *in vivo* experiments all compounds reached maximum intensity within 10 s, but for the mouth model system compounds behaved differently. Low molecular compounds (m/z 69 to 91) reached t_{max} at about 60 s, while higher molecular compounds (m/z 109 to 153) only reached t_{max} after 3 – 5 min. Linear regression of t_{max} as a function of m/z-value resulted in a R^2 of 0.70. Although this is considerably high, it seems more logical from the upper graph in figure 2 to

Table I. Compounds observed with MS-NOSE and identified by MS-MS

M/z	Compound	Observed	Literature[1]
69	unknown	*in vivo, in vitro*	-
73	methylpropanal	*in vivo, in vitro*	(*1, 3, 6, 8, 10, 16*)
75	1-hydroxy-2-propanone (tentative)	*only in vitro*	(*1*)
87	3- and 2-methylbutanal	*in vivo, in vitro*	(*1, 3, 7, 8, 10-12, 16*)
91	glyceraldehyde (tentative)	*only in vitro*	-
95	methylpyrazine	*in vivo, in vitro*	(*1, 3, 5-8, 10, 11, 13*)
109	C2-pyrazine (2,5-dimethyl-, ethyl-)	*in vivo, in vitro*	(*1, 3, 5-8, 10, 11, 13*)
113	2-heptenal	*only in vitro*	(*1, 3, 11-14*)
123	C3-pyrazine (2-ethyl-5-methyl-)	*only in vitro*	(*1, 3, 5-8, 10, 11, 13*)
137	C4-pyrazine (2-ethyl-3,5-dimethyl-, 3-ethyl-2,5-dimethyl-)	*only in vitro*	(*1, 3, 5-7, 10, 11, 13, 16*)
153	2,4-decadienal	*only in vitro*	(*1, 3, 10, 11, 13, 14, 16*)

[1]Previously found in potato or potato products

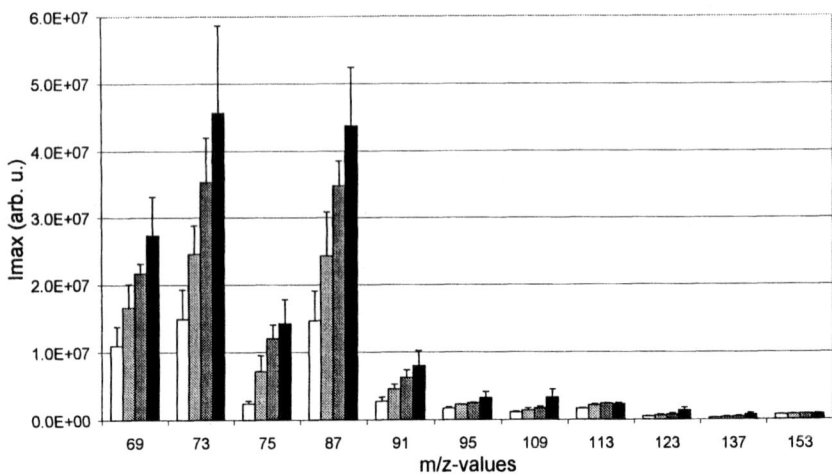

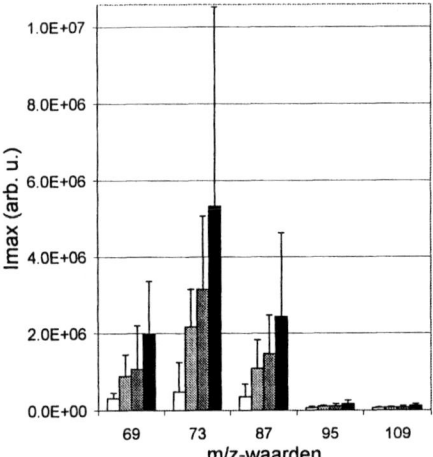

Figure 1. I_{max} in vitro (upper) and in vivo (lower) of compounds released from French fries at 2 min (white), 4 min (light-gray), 6 min (dark-gray), and 8 min (black) of frying at 180°C.

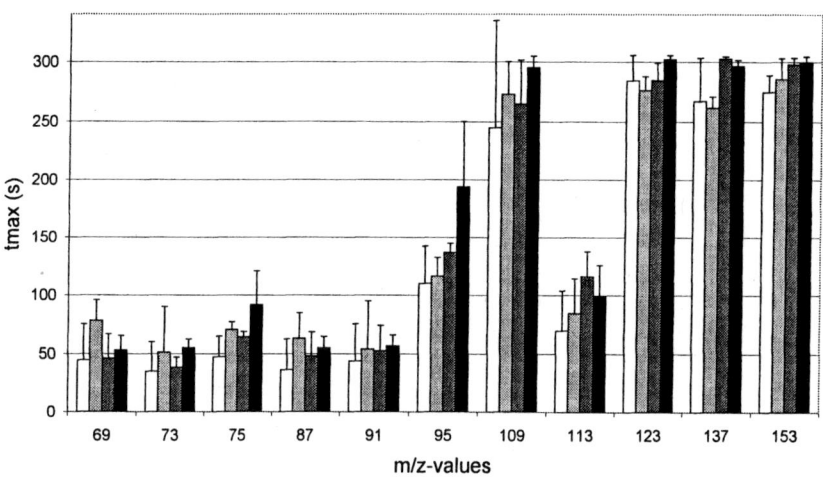
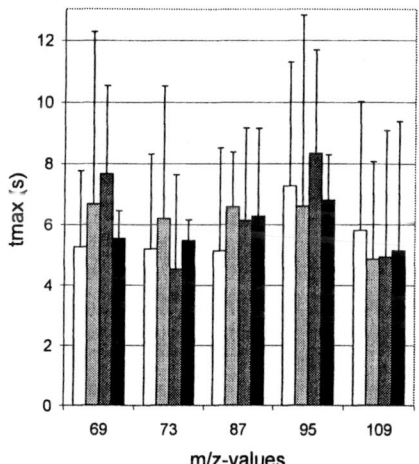

Figure 2. t_{max} in vitro *(upper) and* in vivo *(lower) of compounds released from French fries at 2 min (white), 4 min (light-gray), 6 min (dark-gray), and 8 min (black) of frying at 180°C.*

divide the compounds in groups of fast (m/z 69, 73, 75, 87, 91), intermediate (m/z 95, 113), and slow (m/z 109, 123, 137, 153) release. Even the "fast" compounds *in vitro* are released slowly compared to *in vivo* experiments. Apparently mastication in the mouth model system is far less effective then mastication in the mouth of a panelist.

Effect of salt addition on I_{max} and t_{max}

A trend was observed for both *in vitro* and *in vivo* experiments that I_{max} decreases when salt is added (figure 3), but the effect was not significant. Salt and increased saliva production (*in vivo*) may have an effect on the partition coefficient of some compounds.

There was a significant decrease of t_{max} by addition of salt *in vitro* (figure 4). The effect on t_{max} was not significant for *in vivo* experiments, but showed the same trend. This is in agreement with the effect of frying time; the variation of experiments with panelists was considerably higher than of experiments with the mouth model system. The faster release of compounds after salt addition can be explained by a salting-out effect, causing the concentration of compounds in the vapor phase to increase (*23, 30*).

Conclusions

Using MS-NOSE it is possible to identify and follow the release of flavors from French fries both with panelists and a mouth model system. Maximum intensity of all compounds increased with frying time, and addition of salt caused compounds to release faster. Trends observed in experiments with panelists are confirmed with the mouth model system.

References

1. Maga, J. A., *Food Rev. Int.*, **1994**, *10*: 1-48.
2. Whitfield, F. B.; Last, J.H. In *Volatile compounds in foods and beverages*, Maarse, H., Editor, 1991, Marcel Dekker: New York. p. 222-231.
3. Buttery, R.; Guadagni, D.G.; Ling, L.C. *J. Sci. Food Agric.*, 1973, *24*, 1125-1131.
4. Maga, J. A.; Sizer, C. E. *Lebensmittel Wissenschaft und Technologie*, **1978**, *11*, 181-182.

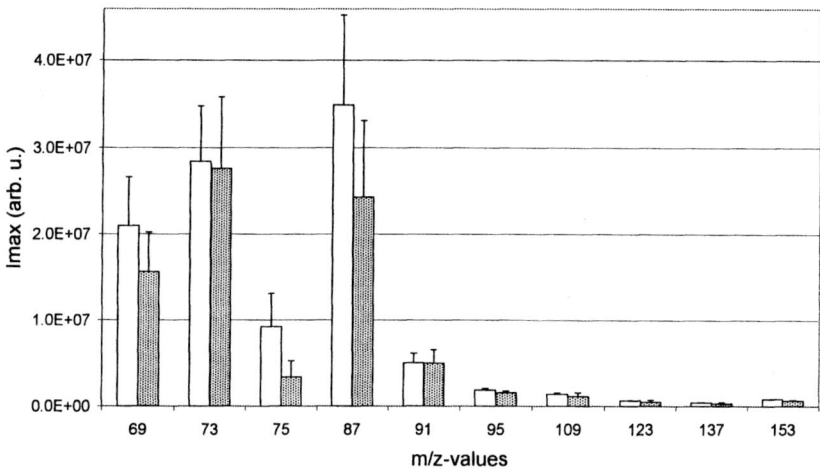

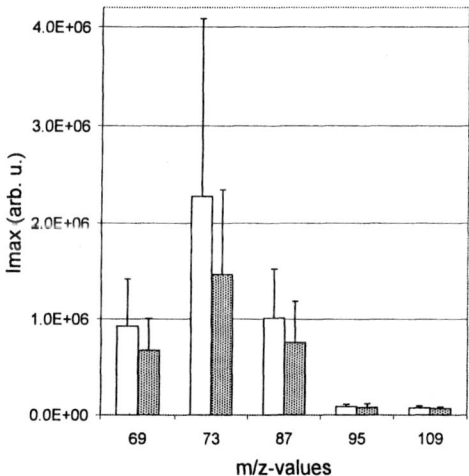

Figure 3. I_{max} in vitro *(upper) and* in vivo *(lower) of compounds released from French fries with salt addition (speckled) and without salt addition (white).*

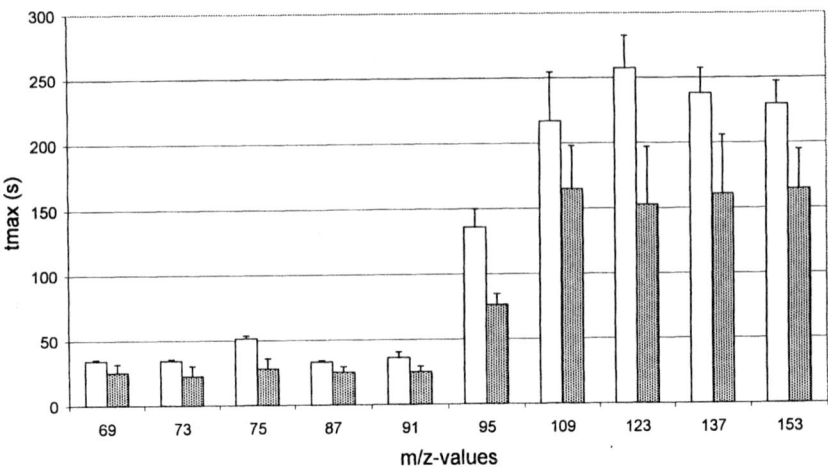
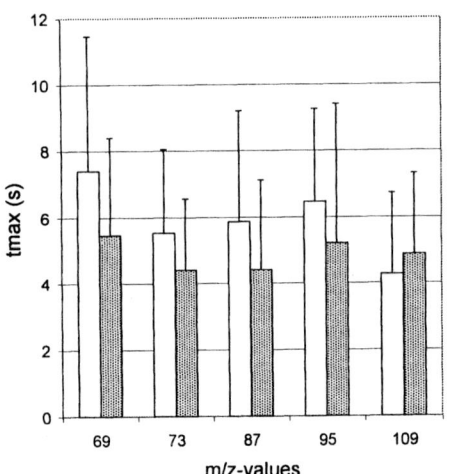

Figure 4. t_{max} in vitro *(upper) and* in vivo *(lower) of compounds released from French fries with salt addition (speckled) and without salt addition (white).*

5. Coleman, E.C.; Ho, C.-T. *J. Agric. Food Chem.*, **1980**, *28*, 66-68.
6. Coleman, E.C.; Ho, C.-T.; Chang, S.S. *J. Agric. Food Chem.*, **1981**, *29*, 42-48.
7. Duckham, S.C.; Dodson, A.T.; Bakker, J.; Ames, J.M. *Nahrung*, **2001**, *45*(5), 317-323.
8. Duckham, S.C.; Dodson, A.T.; Bakker, J.; Ames, J.M. *J. Agric. Food Chem.*, **2002**, *50*, 5640-5648.
9. Ho, C.-T.; Coleman, E.C. *Crit. Rev. Food Sci. Nutr.*, **1980**, *45*, 1094-1095.
10. Martin, F. L.; Ames, J.M. *J. Amer. Oil Chem. Soc.*, **2001**, *78*(8), 863-866.
11. Oruna-Concha, M.J.; Bakker, J.; Ames, J.M. *J. Sci. Food Agric.*, **2002**, *82*, 1080-1087.
12. Oruna-Concha, M.J.; Bakker, J.; Ames, J. M. *Lebensmittel Wissenschaft und Technologie*, **2002**, *35*, 80-86.
13. Oruna-Concha, M.J.; Duckham, S.C.; Ames, J.M. *J. Agric. Food Chem.*, **2001**, *49*, 2414-2421.
14. Petersen, M.A.; Poll, L.; Larsen, L.M. *Food Chem.*, **1998**, *61*(4), 461-466.
15. Carlin, J. T.; Jin, Q.Z.; Huang, T.-C.; Ho, C.-T.; Chang, S.S. *J. Agric. Food Chem.*, **1986**, *34*, 621-623.
16. Wagner, R. K.; Grosch, W. *Lebensmittel Wissenschaft und Technologie*, **1997**, *30*(2), 164-169.
17. Whitfield, F. B. *Crit. Rev. Food Sci. Nutr.*, **1992**, *31*(1/2), 1-58.
18. Wagner, R. K.; Grosch, W. *J. Amer. Oil Chem. Soc.*, **1998**, *75*(10), 1385-1392.
19. Linforth, R. S. T.; Baek, I.; Taylor, A. *J. Food Chem.*, **1999**, *65*, 77-83.
20. Davidson, J. M.; Linforth, R. S. T.; Taylor, A. J. *J. Agric. Food Chem.*, **1998**, *46*, 5210-5214.
21. Cook, D.J.; Linforth, R.S.T.; Taylor, A.J. *J. Agric. Food Chem.*, **2003**, *51*, 3067-3072.
22. Weel, K.G.C.; Boelrijk, E.M.; Alting, A.C.; Mil, P.J.J.M.v.; Burger, J.J.; Gruppen, H.; Voragen, A.G.J.; Smit, G. *J. Agric. Food Chem.*, **2002**, *50*, 5149-5155.
23. Ruth, S.M.v.; Grossmann, I.; Geary, M.; Delahunty, C.M. *J. Agric. Food Chem.*, **2001**, *49*, 2409-2413.
24. Ruth, S. M. v., *Flavour release from dried vegetables*, in *Agrotechnology and Food Sciences*. 1995, Wageningen University: Wageningen. p. 227.
25. Martin, F. L.; Ames, J. M. *J. Agric. Food Chem.*, **2001**, *49*, 3885-3892.
26. Hwang, H.-I.; Hartman, T.G.; Ho, C.-T. *J. Agric. Food Chem.*, **1995**, *43*, 179-184.
27. Brewer, M. S.; Vega, J. D.; Perkins, E.G. *J. Food Lipids*, **1999**, *6*, 47-61.

28. Takeoka, G.; Perrino, C.; Buttery, R. *J. Agric. Food Chem.*, **1996**, *44*, 654-660.
29. Frank, O.; Heuberger, S.; Hofmann, T. *J. Agric. Food Chem.*, **2001**, *49*, 1595-1600.
30. Guichard, E. *Food Rev. Int.*, **2002**, *18*(1), 49-70.

Chapter 5

Flavor Release from Food Emulsions Containing Different Fats

M. Fabre[1], P. Relkin[2], and E. Guichard[1]

[1]UMRA-INRA-ENESAD, 17 rue Sully, B.P. 86510, FR-21065 Djon Cedex, France
[2]Laboratoire de Biophysique, ENSIA, 1 Avenue des Olympiades, FR-91744 Massy, France

Flavor release of aroma compounds from oil-in-water emulsions containing either a vegetable or an animal fat was compared. The fats differ as by their chemical composition, their polarity and by their melting behavior, as determined from DSC experiments. Aroma compounds with different hydrophobicity such as diacetyl, hexenol and ethyl hexanoate were added to the two fats, and the flavor release from emulsified samples was quantified by solid - phase microextraction coupled with GC - MS analysis. The animal or vegetable fats have different effects on flavor release depending on the flavor compounds, except for hexenol which has an intermediate hydrophobicity. The release of the most hydrophilic compound (diacetyl) was higher in bulk and emulsified animal fat (less polar), whereas the release of the most hydrophobic compound (ethyl hexanoate) was higher in bulk and emulsified vegetable fat (more polar).

In multiphasic systems, the partition of the volatile aroma compounds corresponds to the distribution between the lipidic, aqueous and gas phases. Many non-polar volatile molecules being lipophilic, they have a great affinity for lipids, and thus are better retained by them. This is why the majority of aroma compounds have a lower vapor pressure and a sensory threshold higher in oil than in water *(1)*. Moreover, the solubility of volatile compounds in the aqueous and in the oil phases depends on log P values *(2)*. Hydrophobicity plays an important role in the thermodynamic behavior of flavors. The solubility of ethyl acetate and 1-octen-3-ol is higher in organic solvents than in water ; the opposite effect is observed for the more polar 2,5-dimethylpyrazine *(3)*.

Partition of aroma compounds between the lipidic and the vapor phases was studied according to the physical state of the lipid and the nature of the volatile compound. The proportion of solid-to-liquid state of the lipidic phase depends on the nature of the lipids and the temperature of the medium. Maier *(4)* showed that retention of volatile compounds by a triacylglycerol in a solid state increases with the temperature, and the opposite behavior is observed in the liquid state. Indeed, as the temperature increases the solid triacylglycerol becomes liquid and can solubulize aroma compounds, thus a stronger retention of volatile compounds. Small variations in temperature around the melting point of lipids can affect the proportion of solid-to-liquid state. This modification can change the vapor-condensed phase partition coefficient. The sorption of volatile compounds on fatty foods in liquid state is more important than in solid state, due to a better solubility.

In a model cheese, the volatility of diallyl sulfide decreased by 20 % in the presence of tributyrine compared to milk fat *(5)*, whereas diacetyl, which is more hydrophilic, was not affected by the physical state of the fat. This can be explained by the presence of 15 % triacylglycerol in the solid state at 25 °C in the milk fat, which may reduce the solubility of hydrophobic compounds.

The volatility of flavor compounds in oil depends on chain length and degree of unsaturation of fatty acids in the triacylglycerols. The oil-water partition coefficients determined for ethyl butanoate and 2,5-dimethylpyrazine were lower in emulsions with tributyrine (C4) than in those with trioleine (C18:1) *(6)*.

In 1995, Bakker *(7)* observed that the fatty acid composition of fat influenced aroma release. It was shown that stearin (saturated fatty acids) led to a slower flavor release and a lower intensity of the flavor compared to olein (unsaturated fatty acids). In this study, we compared the flavor release of different aroma compounds contained in food model emulsions, according to hydrophobicity and melting behavior of fats.

Experimental

Preparation of the Food Model Emulsions

Emulsions were prepared using 3 % whey protein, 9 % fat, and 0.5 % emulsifier (mixture of mono- and diacylglycerols, guar gum and carrageenan).

Two fats were used, one vegetable fat (a mixture of palm and palmist oils) and one animal fat which is less polar. These fats were provided by Aarhus Oliefabrik A / S and Lactalis industrie, Besnier-Bridel-Aliment, respectively.

The model food emulsions were flavored with three aroma compounds with different hydrophobicities; hexenol, diacetyl and ethyl hexanoate (Table 1), after dissolution in propylene glycol. These compounds were obtained from Food Ingredients Specialities, (FIS, York, England).

Solid - Phase Microextraction (SPME) coupled with GC - MS (SIM mode) analysis

Food model emulsions (5 g) were placed in 20 mL vials and allowed to equilibrate at different temperatures (10, 15, 20 and 25 °C). A SPME fiber, PDMS / DVB, polydimethylsiloxane / divinylbenzene, 65 µm, (Supelco Park, Bellefonte, PA) was used for volatile compounds sampling, using the methodology described (9).

Volatile compounds were desorbed by inserting the fiber into the GC injector, set at 250 °C for 10 min, 1 min for desorption (purge off) and 9 min for cleaning (purge on). All the SPME operations were automated using a MPS2 MultiPurpose Sampler (Gerstel, Applications, Brielle, The Netherlands). Experiments were performed in triplicate.

GC-MS analysis

A HP 6890-GC equipped with a split / splitless injector coupled with a mass selective detector 5970 (Hewlett Packard, Palo Alto, CA) was used. A fused-silica capillary column DB-Wax, 50 m, 0.32 mm ID, 1 µm film thickness (J & W Scientific, Agilent Technologies, Folsom, CA) was employed. The carrier gas was helium (35 cm.s^{-1}).

The GC oven heating was started at 50 °C, then increased to 220 °C at a rate of 5 °C. min^{-1}. The total analysis time was 39 min. The injector was maintained at 250 °C.

The mass spectrometer was operated in the mass range from 29 to 300 at a scan rate of 1.89 sec.scan^{-1}. The quantification was realized using SIM mode (Selective Ion Monitoring). The selected and specific ions were 43 for diacetyl, 101 for ethyl hexanoate and 82 for hexenol.

Determination of vapor - oil partition coefficients

The studied fats were solid at room temperature. To introduce aroma compounds, it was necessary to melt fats at 40 °C for 15 min. Aroma concentrations varied according to aroma compounds (200, 400 and 2000 ppm, respectively, for diacetyl, ethyl hexanoate and hexenol).

Samples (5 g) were placed directly on the tray of the automatic injector and allowed to equilibrate at 10, 20 and 40 °C for 150 min. One milliliter of the gas phase was injected in the gas chromatograph with the same procedure used for solid – phase microextraction experiments. Experiments were done in triplicate.

Vapor - oil partition coefficient was equal to aroma concentration in vapor phase divided by aroma concentration in fat.

$$K_i = \frac{y_i}{x_i}$$

With x_i = aroma (i) concentration in fat phase (mg.kg^{-1}) and y_i = aroma (i) concentration in vapor phase (mg.kg^{-1}).

Determination of melting behavior of fats and food model emulsions by Differential Scanning Calorimetry (DSC)

Differential scanning calorimetry (DSC, Perkin-Elmer 7 - Software Pyris, Norwalk, US) was used to study the melting behavior of pure fats and food model emulsions, as previously described *(10)*. After storage at – 30 °C, the samples (around 15 mg) were put in an aluminium pan at room temperature, with an empty pan used as a reference pan. In first set of experiments, the samples were heated from – 30 °C (bulk fat) or – 10 °C (food model emulsions) to 50 °C, at 5°C.min^{-1}, and the heat of melting ΔH_0, was calculated from the area under the peak transitions. In second set of experiments, each sample was held for 30 min at the temperatures ranging from 5 to 25 °C, and then heated to 50 °C, at 5 °C.min^{-1}. The percentage (%) of liquid was deduced from the following equation.
% liquid = 100 * ($\Delta H_0 - \Delta H$) / ΔH_0,
where ΔH is the heat of reaction of the annealed samples.

Results and Discussion

Flavor release analysis by solid – phase microextraction

Solid phase microextraction (SPME), coupled with GC-MS (SIM mode) was used to quantify aroma release in the vapor phase (Figures 1, 2 and 3) . The polydimethylsiloxane/divinylbenzene (PDMS / DVB) coating fiber was chosen because its sensitivity was adapted to the three aroma compounds *(11)*. The experiments were done at 10, 15, 20 and 25 °C, in triplicate.

The flavor release of the three aroma compounds was different according to the nature of fat. The nature of fat seemed to have more influence on the release of diacetyl and ethyl hexanoate than hexenol.

The release of hydrophobic compounds such as ethyl hexanoate (log P = 2.80) was more important in emulsion with the vegetable fat used,

Table I. Aroma Compounds: Odor Threshold, Odor Description, Log P and Concentration in the Emulsion.

Aroma compounds	Odor threshold, mg.kg^{-1} in water (literature cited)	Odor description	Log P (2)	Concentration (mg. kg^{-1})
Ethyl hexanoate	0.0003 (8)	Fruity	2.80	2.80
Hexenol	0.5 (8)	Green	1.40	1.60
Diacetyl	0.003 (8)	Buttery	-2.26	0.56

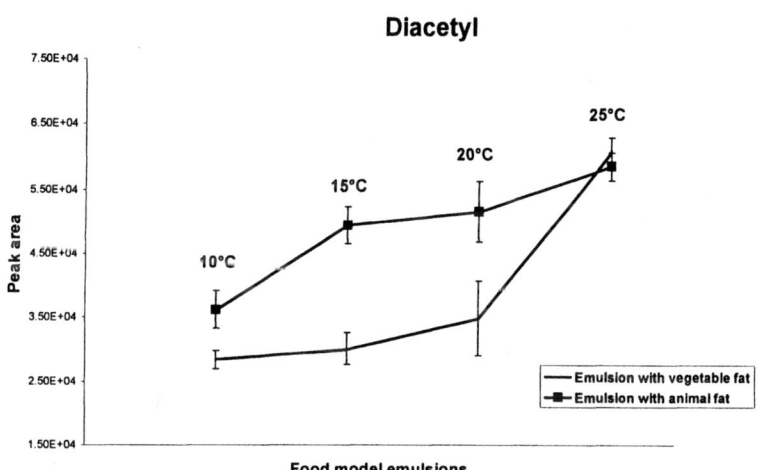

Figure 1. Flavor release of diacetyl from two different food model emulsions by solid-phase microextraction (SPME-GC-MS) according to temperatures.

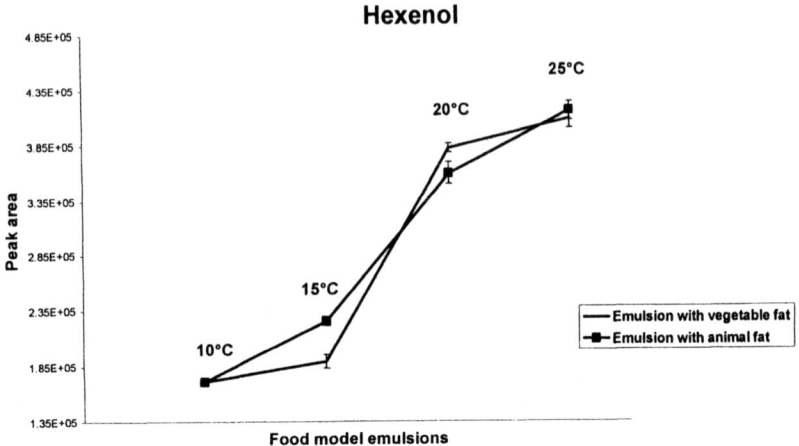

Figure 2. Flavor release of hexenol from two different food model emulsions by solid-phase microextraction (SPME-GC-MS) according to temperatures.

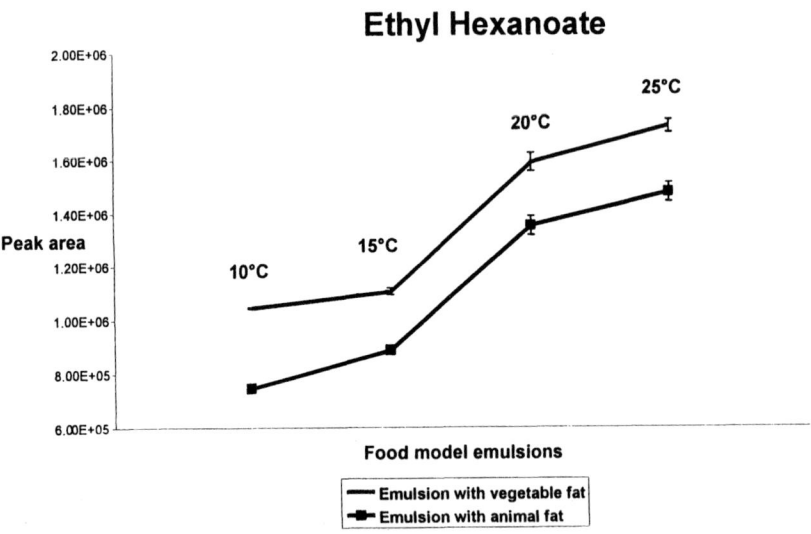

Figure 3. Flavor release of ethyl hexanoate from two different food model emulsions by solid-phase microextraction (spme-gc-ms) according to temperatures.

and was less important in emulsion which contained the animal fat. The opposite was observed for the release of a hydrophilic compound such as diacetyl (log P = -2.26). The flavor release of hexenol, which had an intermediate log P value (log P = 1.40), did not seem to be influenced by the nature of fat. Thus, hydrophobicity of aroma compounds was an important factor that needed to be taken into account in order to explain flavor release from food model emulsions. Moreover, it was necessary to rationalize the differences in flavor release from different oils, and thus vapor - oil partition coefficients for each aroma compound at different temperatures were determined.

Vapor-oil partition coefficients

The results of vapor-oil partition coefficients determined at different temperatures (10, 20 and 40°C) for each aroma compound are reported in Table II. The more the temperature increased, the more the value of vapor - oil partition coefficient increased, which indicates that flavor release was more important when temperature was higher. However, differences were more important between 20 and 40°C than between 10 and 20°C. This could be explained by the proportion of solid and liquid phases in the fats which is dependent on the temperature. At a lower temperature, the percentage of solid fat was more important and thus fat could not solubilize aroma compounds any more.

At a temperature of 40 °C, the two fats are in the liquid state and thus differences in vapor - oil partition coefficients could be explained by differences in affinity between fat and aroma compounds. This effect of the nature of fat also depends on the nature of the aroma compound.

Diacetyl, the most hydrophilic aroma compound, was the only one which had a higher vapor - fat partition coefficient in the animal fat than in the vegetable fat. Animal fat was more apolar than vegetable fat in our study. This was because animal fat contains a greater proportion of triacylglycerols (Table III).

Vapor - oil partition coefficients of hexenol and ethyl hexanoate did not significantly change for the two different fats examined. Thus, they did not seem to be influenced by the nature of the fat.

These experiments indicate that the hydrophobicity of aroma compounds' and the nature of fat could influence flavor release. To complement our study, the melting behavior of pure fats and food model emulsions was also studied using differential scanning calorimetry (DSC).

Determination of the melting behavior of pure fats and food model emulsions by differential scanning calorimetry (DSC)

The evolutions of liquid fat following melting the two fat samples used for emulsion preparation are shown in Figure 4. At 35°C, the two fats were in 100 % liquid state. However, the proportion of liquid seemed to be higher for

Table II. Vapor – Fat Partition Coefficients (Animal And Vegetable Fats) at Different Temperatures for the Three Aroma Compounds.

Aroma compounds; °C	Vapor - animal fat partition coefficient ($*10E^{-6}$)	Vapor – vegetable fat partition coefficient ($*10E^{-6}$)
Diacetyl		
10	1490.0	1240.0
20	2550.0	2530.0
40	5140.0	4080.0
Hexenol		
10	5.0	7.7
20	8.5	9.7
40	24.0	24.0
Ethyl hexanoate		
10	20.0	25.0
20	24.0	28.0
40	80.0	81.0

Table III. Detemination of Proportions of Di and Triacylglycerols in the Pure Fats by Iatroscan.

Fats	*Diacylglycerols (%)*	*Triacylglycerols (%)*
Animal fat	1.7	98.3
Vegetable fat	4.8	95.2

animal fat than for the vegetable fat, for holding temperatures ranging from 10 to 30°C. This result could be explained on the basis of the triacylglycerol composition of the two fats. The animal fat having a higher triacylglycerol content (Table III) was more apolar, and consequently melted at a lower temperature than the vegetable fat which was more polar.

The melting behavior of the food model emulsions is shown in Figure 5. Probably nonemulsified fat samples, food model emulsion containing animal fat, presented a higher proportion of liquid than emulsion containing vegetable fat, at temperatures ranging from 15 to 25°C. Figures 4 and 5 show that 50 % liquid was formed at around 18°C for the animal fat and at 23°C for the vegetable fat. At 25°C, the temperature of solid-phase microextraction analysis, the proportion of liquid for animal fat was lower (65%) in the food model emulsion than in the bulk sample (75%), whereas this difference was slightly lower for vegetable fat. Thus, these experiments indicated differences between the melting behavior of the two fats both in bulk and in the emulsified forms.

Discussion

The flavor release by solid phase microextraction for three different aroma compounds from two emulsions which differed by the nature of fat was studied. The three aroma compounds differed by their hydrophobicity and the two fats by their polarity and thus by their melting behavior. Diacetyl, the only hydrophilic aroma compound used was more soluble in water than in fat *(12)*. It was also the only one which was more released fom emulsion which contained animal fat. The apolar nature of animal fat and its high triacylglycerol content could explain its higher proportion of liquid and the data of vapor - oil partition coefficient, as a function of temperature. The results are in agreement with vapor - oil partition coefficients reported by Overbosch *et al.* *(13)* which were 2.4 x 10^{-3} (20°C) and 3.7 x 10^{-3} (37°C), but slightly different from those obtained by Salvador *et al.* *(14)* of 6 x 10^{-4} in sunflower oil (25°C).

The experiments on emulsions by solid-phase microextraction and determination of vapor - oil partition coefficient showed that hexenol was not influenced by the nature of fat. This aroma compound had an "intermediate" hydrophobicity. These results are in agreement whith those of Miettinen *et al.* *(15)*. These authors studied ice cream which contained 9 % animal fat or 9 % vegetable fat. No significant difference in flavor release of hexenol was observed.

Ethyl hexanoate was the most hydophobic aroma compound in this study. It was released most from emulsion which contained the vegetable fat. This fat which contained less triacylglycerols was more polar than the animal fat and it melted at a higher temperature (DSC) than the animal fat. However, no significant differences were observed in the vapor - fat partition coefficients between animal fat and vegetable fat at 40 °C. The results are in agreement with those reported by Landy *et al.* *(16)*, in triolein at 25°C, which was 3 x 10^{-5}.

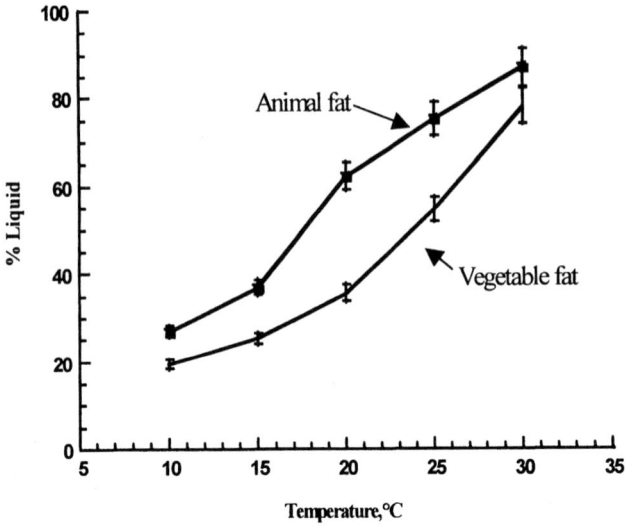

Figure 4. Evolution of liquid- to- solid proportion in bulk animal fat and in bulk vegetable fat, as determined by differential scanning calorimetry (DSC).

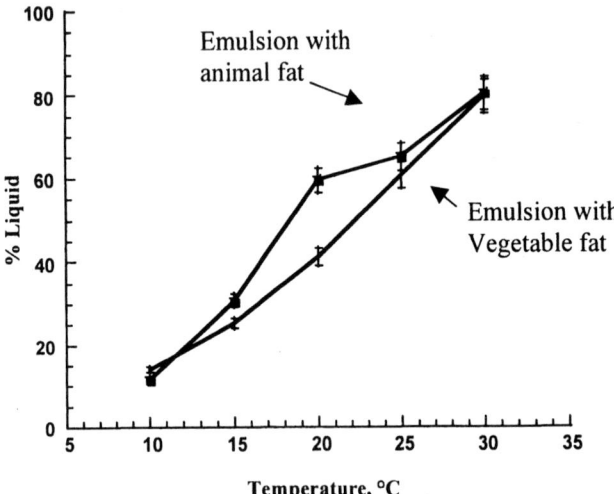

Figure 5. Evolution of liquid- to- solid proportion of food model emulsions containing animal fat and vegetable fat, as determined by differential scanning calorimetry (DSC).

Charles *et al. (17)* showed that flavor release of ethyl hexanoate in emulsion was influenced by the structure of the emulsion and the nature of the interface. In this study, as the emulsions had the same average droplet size (0.6 – 0.8 µm), it is possible that the nature of the interface varied (amount of adsorbed proteins, changes in conformation of the adsorbed proteins) as already suggested by Rogacheva *et al. (18)*. Thus, the flavor release appears to be influenced not only by the melting behavior of fat but also by the nature of aroma compounds.

Acknowledgment

This work was financially supported by the Nestlé company and the Regional Council of Burgundy. We thank Fis (York, UK) for providing aroma compounds and flavor mixture, and Aurélien Neveu (technician at ENSIA) for his help in DSC measurements.

References

1. Bakker, J.; Brown, W.; Hills, B.; Boudaud, N.; Wilson, C.; Harrison, M. In *Flavour Science Recent Developments. 8th Weurman Flavour Research Symposium*; A. J. Taylor and D. S. Mottram, Eds.; The Royal Society of Chemistry: Reading, 1996; pp 369-374.
2. Rekker, R.F. In *Pharmacochemistry Library;* W.Nauta and R.F. Rekker, Eds.; Elsevier Scientific: Amsterdam, 1977.
3. Druaux, C. ; Le Thanh, M.; Seuvre, A-M.; Voilley, A. *J. Am. Oil Chem. Soc.* **1998**, *75*, 441–445.
4. Maier, H.G. In *Aroma Research*; H. Maarse, P.J. Groenen, Eds.; Pudoc: Wageningen, 1975; pp. 143-157.
5. Dubois, C.; Sergent, M.; Voilley, A. In *Flavor – food Interactions*; RG McGorrin, JV Leland, EDS.; American Chemical Society: Washington, D.C. 1996; pp 217 – 226.
6. Harvey, B.; Druaux, C.; Voilley, A. In *Food Macromolecules and Colloids*; E Dickinson, D Lorient, Eds.; The Royal society of chemistry: Cambridge, 1995; pp 154-163.
7. Bakker, J. In *Ingredient interactions. Effect of food quality.* A.G. Gaonkar, Ed., Marcel Dekker Inc.: New-York, 1995; pp 411-439.
8. Pyysalo, T.; Suiko, M.; Honkanen, E. *Lebensm. – Wiss. Technol.* **1977**, *10*, 36-39.
9. Fabre, M.; Aubry, V.; Guichard, E. *J. Agric. Food Chem.* **2002**, *50*, 1497-1501.
10. Relkin, P.; Talleb, A-A.; Sourdet, S.; Fosseux, P-Y. *J. Am. Oil Chem. Soc.* **2003**, *80*, 741-746.

11. Roberts, D.D.; Pollien, P.; Milo, C. *J. Agric. Food Chem.* **2000**, *48*, 2430-2437.
12. Land, D.G.; Reynolds, J. In *3th Weurman Flavour Research Symposium,* P. Schreier, Ed.; Walter de Gruyter: Berlin, 1981; 3, pp 701 – 705.
13. Overbosch, P.; Afterhof, W.G.M.; Haring, P.G.M. *Food Rev. Inter.* **1991**, *7 (2),* 137–184.
14. Salvador, D.; Bakker, J.; Langley, K.R.; Potjewidj, R.; Martin, A.; Elmore, S. *Food Qual. Prefer.* **1994**, *5*, 103-107.
15. Miettinen, S.M.; Tuorila, H.; Piironen, V.; Vehkalahti, K.; Hyvonen, L. *J. Agric. Food Chem.* **2002**, *50*, 4232–4339.
16. Landy, P.; Courthaudon, J.L.; Dubois, C.; Voilley, A. *J. Agric. Food Chem.* **1996**, *44,* 526–530.
17. Charles, M.; Lambert, S.; Brondeur, P.; Courthaudon, J.L.; Guichard, E. In *Flavor Release,* American Chemical Society, DD Roberts and AJ Taylor (Eds). 2000; 763, pp 342 – 354.
18. Rogacheva, S.; Espinoza – Diaz, M.; Voilley, A. *J. Agric. Food Chem.* **1999**, *47*, 259–263.

Chapter 6

Changes in Key Odorants of Sheep Meat Induced by Cooking

Valerie Rota and Peter Schieberle

Deutsche Forschungsanstalt für Lebensmittelchemie, Lichtenbergstrasse 4, D–85748 Garching, Germany

Application of the Aroma Extract Dilution Analysis on two extracts prepared from cooked and raw lean sheep meat revealed 4-ethyloctanoic acid (mutton-like), trans 4,5-epoxy-(E)-2-decenal (metallic), (Z)-1,5-octadien-3-one (geranium-like) and (E,E)-2,4-decadienal (deep fried) as important odorants in the cooked as well as in the raw meat, thereby indicating the important role of the raw meat as source of sheep meat odorants. 4-Hydroxy-2,5-dimethyl-3(2H)-furanone and 2-acetyl-1-pyrroline as well as 2-aminoacetophenone were among the few aroma compounds, which were clearly increased in their Flavor Dilution (FD) factors by the cooking procedure. A stable isotope dilution assay showed that 4-ethyloctanoic acid was by a factor of 4 higher in uncooked intramuscular fat as compared to uncooked adipose tissue, in which 4,5-epoxy-(E)-2-decenal and further lipid degradation were shown to be the most odor-active compounds. The cooking procedure did not much alter the concentration of 4-ethyloctanoic acid.

As compared to pork or beef meat, the consumption of sheep meat (mutton or lamb) is on a low level in most of the European countries. For example in Germany, sheep meat accounted for only 1.3 % of the total annual meat used in Germany. The main reason for this is undoubtedly the characteristic sweet-fatty odor developing during cooking of sheep meat, which is regarded as unpleasant and, thus, leads to a negative consumer appraisal. In search of the compounds responsible for this characteristic aroma, Hornstein and Crowe (1) were the first to report on the key role of adipose tissue as source of the species-specific aroma compounds. Because of the intense odor, Wong *et al.* (*2*) later suggested branched chain fatty acids, in particular, 4-methyloctanoic and 4-methylnonanoic acid as being responsible for the mutton-like aroma of cooked sheep meat. In a study on the basic volatiles of roasted lamb fat, Buttery *et al.* (*3*) identified 2-pentylpyridine which, based on its unpleasant odor at a threshold in water of 0.6 µg/L, was suggested as another cause for the consumer rejection of cooked sheep meat. Ha and Lindsay (*4*) suggested another branched fatty acid, 4-ethyloctanoic acid, as a further odorant contributing to the typical aroma developing in mutton meat during cooking. The same group (*5*) later proposed several alkylphenols, which were present above their odor thresholds in perinephric mutton fat, as additional odorants possibly responsible for the sheepy-muttony aroma of sheep meat.

Guided by the first report of Hornstein and Crowe (*1*), most of the studies on sheep meat aroma were focused on the formation of such "muttony" smelling odorants generated from adipose tissue. Sutherland and Ames (*6*) and Elmore *et al.* (*7*) were among the first to characterize the entire set of volatiles formed from either lean lamb meat or adipose tissue. Sutherland and Ames (*6*) identified 132 volatile compounds in pan-fried lamb, most of them were assigned as lipid degradation products, such as octanal. 4,6-Dimethyl-1,3-oxathione eliciting a "stale-wet animal" odor note at the odor port was particularly mentioned as new constituent of sheep meat by these authors (*6*). Elmore *et al.* (*7*) compared the volatiles formed by pressure-cooking of lean lamb meat as effected by diet and breed. They reported a significantly higher amount of lipid oxidation products in animals fed fish oil, whereas the breed influenced the concentration of certain compounds derived from Maillard-type reactions.

It is now well accepted in the literature that only those volatiles exceeding their odor threshold can be suggested as contributors to food aromas (*8*). Such aroma compounds can be selected from the bulk of less odor-active or odorless volatiles by application of dilution to odor threshold techniques, such as on Aroma Extract Dilution Analysis (AEDA) followed by a calculation of odor activity values (*8*). However, except from one study focused on the effect of stress on sheep meat (*9*), using GC/Olfactometry, such data are scarcely available. Furthermore, because no studies have been performed up to now to elucidate differences in the aroma compounds before and after cooking of sheep meat, it is quite unclear which volatiles are already present in the raw sheep meat and which are formed from odorless precursors during processing. Using AEDA and quantitative measurements, the purpose of the following study was to close this gap by characterizing the key aroma compounds in raw and cooked sheep meat based on the combination of sensory experiments with analytical data.

Materials and Methods

Meat from female sheep carcasses (about 6 years old animals) was used in the study. The carcasses were purchased from a company specialized in sheep trade and sale (Bayerlamm, Munich, Germany). Adipose tissue was removed and aliquots of about 500 g of lean meat were frozen at –60 °C prior to cooking. Total storage time was about 4 months. For aroma analysis, the meat was cut into large pieces (300 g; 3 cm x 3 cm) and, after addition of tap water (300 mL), pressure-cooked for 25 min in a kitchen cooker. The broth was separated, the pieces of meat were frozen in liquid nitrogen and, after homogenization, the volatile fraction was isolated by extraction with dichloromethane followed by high vacuum distillation as recently described (10). Aroma volatiles, present in raw sheep meat from the same carcasse were isolated based on the same amount of meat. Quantitation of deleted volatiles was done by stable isotope dilution assays as recently reported (10).

Results and Discussion

Key odorants in cooked lean sheep meat

In general, lamb meat is rated superior in flavor as compared to the meat of older animals. However, it is not clear from literature data, which flavor compounds are different depending on age. In order to get more information on the flavor compounds of elder animals, we used the meat of six years old sheep, which are normally at the borderline for consumption. Nevertheless, the meat was palatable and did not show any "off-note", but elicited an intense overall aroma, in which meat-like, fatty, sweet caramel-like and roasty odor notes prevailed. Interestingly, the overall aroma of the meat did clearly resemble the aroma of cooked beef, whereas the typical "mutton-like" aroma was rated low by the sensory panel. To evaluate the odorants showing the highest Flavor Dilution (FD) factors, an AEDA was performed (8). For this purpose, the volatile fraction was treated with sodium bicarbonate prior to the AEDA in order to separate the neutral/basic from the acidic volatiles (10). In Figure 1, the Flavor Dilution chromatogram obtained by application of the AEDA on the neutral/basic fraction is displayed. A total of fourty-six odor-active areas was detectable, among which nos. 40 (metallic), 17 (mushroom-like), 18 (geranium-like), 30 (green, cucumber-like), 31 (green, tallowy) and 38 (fatty, deep fried) showed the highest FD-factors in the range of 1024 to 2048. In the acidic fraction, another twenty-five odor-active volatiles were detected, among which two compounds exhibited the highest FD-factors, namely no. 10a (caramel-like) and no. 16a (sheep-like).

In order to identify the compounds responsible for these odor impressions, a distillate, prepared from 1 kg of cooked lean meat, was separated by column chromatography on silica gel (10) and mass spectra were recorded from the respective GC volumes, in which an odor was perceivable. Based on the use of reference compounds, the eight aroma compounds displayed in Figure 2, showing the highest FD-factors in the cooked meat, were identified. The identification experiments in combination with the FD-factors revealed 4-ethyloctanoic acid (no.16a), 4,5-epoxy-(E)-2-decenal (no. 40), 4-hydroxy-2,5-dimethyl-3(2H) furanone (no. 10a), 1-octen-3-one (no. 17), (Z)-1,5-octadien-3-one (no. 18), (E,Z)-2,6-nonadienal (no. 30), (E)-2-nonenal (no. 31) and (E,E)-2,4 decadienal (no. 38) as key odorants in the cooked meat. Four of these odorants, labeled with an asterisk in Figure 2, are reported for the first time as sheep meat constituents. The results suggested that among the branched chain fatty acids reported as "mutton-like" odorants in sheep meat in the literature, 4-ethyloctanoic acid is the most odor-active.

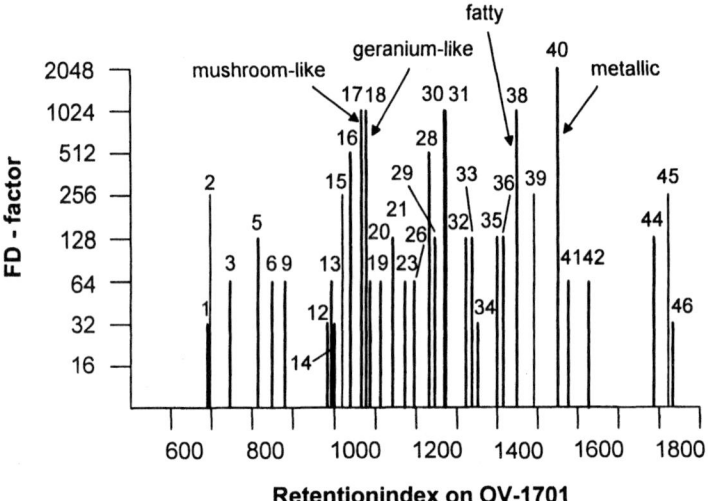

Figure 1. Flavor Dilution chromatogram of odor-active, neutral/basic volatiles isolated from cooked sheep meat

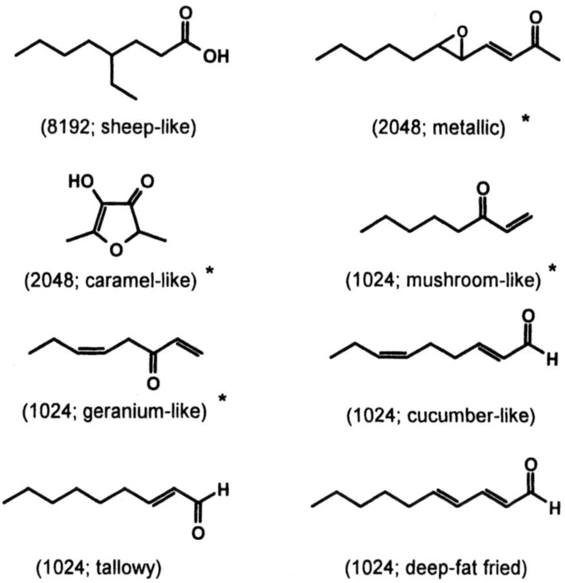

*Figure 2. Most important aroma compounds identified in cooked sheep meat
(*: Previously not reported as sheep meat constituent)*

Key odorants in raw lean sheep meat

The key odorants in raw lean sheep meat were identified by application of the AEDA on an extract prepared from minced, raw meat followed by identification experiments using reference compounds (*10*). Identifications were focused on those volatiles showing FD-factors in the range of 4 to 8192. In Table I, a selection of odorants either detected with FD-factors ≥ 256 in the raw meat or those showing significant differences as compared to the cooked meat are displayed. The complete set of odorants is given in (10). The highest FD factor was found for 4-ethyloctanoic acid followed by the seasoning-like smelling 3-hydroxy-4,5-dimethyl 2(5H)-furanone, the metallic trans-4,5-epoxy-(E)-2-decenal, as well as (Z)-1,5-octadien-3-one and (E,E)-2,4-decadienal eliciting geranium-like and deep fried odor qualities.

Table I. Selected key odorants identified in raw lean sheep meat. Comparison with cooked meat.

Odorants	Odor quality	RI on FFAP	FD in raw	FD in cooked meat
4-Ethyloctanoic acid	mutton-like	2192	2048	8192
3-Hydroxy-4,5-dimethyl-2(5H)furanone	seasoning-like	2188	1028	256
trans-4,5-Epoxy-(E)-2-decenal	metallic	2048	1024	2048
(Z)-1,5-Octadien-3-one	geranium-like	1377	512	1024
(E,E)-2,4-Decadienal	deep-fried	1815	512	1024
(E,E)-2,4-Nonadienal	deep-fried	1707	256	128
(E,Z)-2,4-Nonadienal	tallowy, fatty	1648	256	128
(E)-2-Nonenal	green/tallowy	1535	256	1024
1-Octen-3-one	mushroom-like	1295	256	1024
4-Hydroxy-2,5-dimethyl-3(2H)-furanone	caramel-like	2035	<1	2048
4-Methyloctanoic acid	muttony	2094	64	512
Butanoic acid	rancid	1610	64	256
2-Acetyl-2-thiazoline	roasty	1743	64	512
2-Aminoacetophenone	medicinal	2210	4	256
Methional	potato-like	1439	128	512
2-Acetyl-1-pyrroline	roasty	1330	16	256
2,3-Butandione	buttery	985	128	256

See ref (*10*).

A comparison of FD-factors of the nine key odorants in raw meat (FD ≥256) with those determined for the respective odorants in the cooked meat clearly showed that, e.g., the mutton-like smelling 4-ethyloctanoic as well as several compounds known as lipid oxidation products, such as 4,5-epoxy-(4)-2-decenal, (E,E)-2,4-decadienal or (E,E)-2,4-nonadienal were already present with similar intensities in the raw sheep meat. Because these compounds are known as degradation products of hydroperoxides formed during an oxidation of unsaturated fatty acids (cf. Figure 3), the data suggest that these compound may be formed during storage of the raw meat rather than by oxidative processes during cooking.

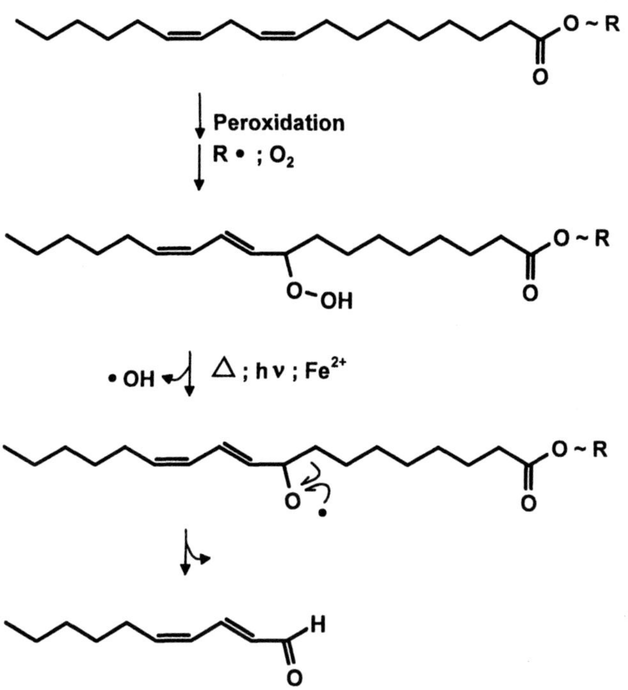

Figure 3. Scheme for the generation of (E,Z)-2,4-decadienal by a peroxidation of linoleic acid via 9-hydroperoxy-octadecadienoate

On the other hand, the cooking process only enhanced the FD factors of a few odorants, in particular, 4-hydroxy-2,5-dimethyl-3(2H)-furanone, 2-aminoacetophenone and 2-acetyl-1-pyrroline. The latter is a well-known degradation products of the amino acid proline.

Quantitative measurements

Flavor dilution factors are only roughly correlated with concentrations, because different work-up losses between different samples are not taken into account. To confirm the results on the presence of odorants in the raw meat and, also, to establish the formation of some compounds from precursors, quantitative experiments using isotopically labeled internal standards were performed. Details on the synthesis of the standards and method development are reported elsewhere (10). The results showed (Table II) that 4-ethyloctanoic acid did not differ significantly between the raw and the cooked sample and also 4-methyloctanioc acid as well as the aldehydes (E)-2-nonenal, (E,E)-2,4-decadienal and (E,E)-2,4-nonadienal were only increased by a factor of 2 during cooking. A comparison of the concentrations of selected lipid derived odorants in sheep meat with data previously reported in cooked beef and pork meat revealed that the differences in the concentrations were not very pronounced (Table III). These data suggest that also in pork and beef such lipid derived aldehydes are already present in the raw materials, however, no such studies on raw pork or beef meat have yet been performed. On the other hand, in particular, 4-hydroxy-2,5-dimethyl-3(2H)-furanone was significantly enhanced in the cooked sample. The furanone is known to be formed from fructose-1,6 biphosphate (11), a precursor which has previously been identified in, e.g. extracts of raw beef meat (12).

Table II. Comparison of the concentrations of selected aroma compounds in raw (RSM) and cooked (CSM) lean sheep meat

Odorant	Conc. (µg/kg)	
	RSM	CSM
4-Ethyloctanoic acid	255	217
4-Methyloctanoic acid	278	502
(E)-2-Nonenal	27	21
(E,E)-2,4-Decadienal	2.9	4.6
(E,E)-2,4-Nonadienal	1.4	3.8
(Z)-1,5-Octadien-3-one	0.8	2.1
4-Hydroxy-2,5-dimethyl-3(2H) furanone	<50	9162
2-Acetyl-1-pyrroline	0.08	0.2

Table III. Comparison of the concentrations (µg/kg) of selected odor-active oxo-compounds in different types of cooked meat

Aroma compound	Sheep	Beef[a]	Pork[a]
Hexanal	585	345	173
(E)-2-Nonenal	21	32	15
(Z)-2-Nonenal	0.9	6.2	1.4
(E,Z)-2,6-Nonadienal	3.4	1.5	n.d.
(E,E)-2,4-Nonadienal	3.8	n.d.	2.0
(E,E)-2,4-Decadienal	15	27	7.4
(E,Z)-2,4-Decadienal	4.6	n.d.	7.2
1-Octen-3-one	2.6	9.4	4.8

[a] Data from Kerscher and Grosch (13). n.d.: not determined

These results clearly indicated that the free methyl branched fatty acids are already present as aroma compounds in the raw lean meat, because literature data often report on the total fatty acids liberated by hydrolysis prior to the analysis.

As explained in the Introduction, literature results have pointed out that 4-methyl- and 4-ethyloctanoic acid are important aroma constituents of sheep meat and stem from adipose tissue. Because the lean meat used in our studies only contained the intramuscular fat, it was of interest to compare the amounts of the free branched chain fatty acids in the adipose tissue and the intramuscular fat from the same animal. For this purpose, the intramuscular fat was isolated by a well-known extraction method (*10*) and the amounts of free methylbranched fatty acids were quantified. The results (Table IV) showed that the amounts of 4-methyl- and, particularly, 4-ethyloctanoic acid were clearly higher in the uncooked intramuscular fat as compared to the adipose tissue. These data explain why the characteristic odor of cooked sheep meat caused by these acids cannot be completely avoided by cutting-off the tissue fat before cooking.

Table IV. Comparison of the amounts of free 4-methyl- and 4-ethyloctanoic acid in uncooked adipose tissue (AT) and uncooked intramuscular fat (IF) of sheep meat

Fatty acid	Conc. ($\mu g/kg$)	
	AT	IF
4-Methyloctanoic acid	1154	3279
4-Ethyloctanoic acid	306	1487

Key odorants in sheep tissue fat

In the carefully melted state, sheep adipose tissue fat as such, i.e. without applying heat, has a very intense fatty, soapy overall aroma. To elucidate the most odor-active compounds in the unprocessed fat, this was dissolved in diethyl ether and the volatiles were isolated by high vacuum distillation. Among the fourty-five odorants detected (*10*), trans-4,5-epoxy-(E)-2-decenal (metallic), (E,E)-2,4-decadienal (deep fried), (E)-2-nonenal (green, tallowy), (Z)-2-nonenal and (E)-2-decenal were the most odor-active (Table V). Currently, quantitative studies are undertaken to determine their role in the overall aroma based on odor activity values.

In summary, these results corroborated that significant amounts of lipid oxidation products are already present in the raw sheep meat (lean muscle and adipose tissue) and, consequently, these are also aroma compounds of processed meat or meat products. Our data, however, suggest that their formation by a lipid oxidation during meat processing seems to be of minor importance. Further studies will be performed to corroborate this assumption by quantitative measurements. Although the data seem to propose that the aromas of raw and cooked meat are identical, it has to be pointed out that the texture of raw and cooked meat is completely different and may, thus influence the release of certain compounds. To characterize the compounds responsible for the aromas of raw and cooked meat flavor recombination experiments will be performed. Furthermore, more studies are necessary to clarify the influence of age on the aroma compounds of sheep meat.

Table V. Key aroma compounds (FD ≥256) identified in uncooked adipose tissue from sheep meat

Odorant	Odor quality	FD-factor
trans 4,5-Epoxy-(E)-2-decenal	metallic	4096
(E,E)-2,4-Decadienal	deep fried	4096
(E)-2-Nonenal	green, tallowy	4096
(Z)-2-Nonenal	tallowy	2048
(E)-2-Decenal	green, fatty	1024
(E,Z)-2,4-Decadienal	tallowy, fatty	512
1-Octen-3-one	mushroom-like	512
(Z)-1,5-Octadien-3-one	geranium-like	512
Unknown	butter-like, roasted	256
(E,Z)-2,6-Nonadienal	cucumber-like	256
(Z)-2-Decenal	green, fatty	256

References

1. Hornstein, J.; Crowe, P.F. *J. Agric. Food Chem.* **1963**, *11*, 147-149.
2. Wong, E.; Nixon, L.N.; Johnson, C.B. *J. Agric. Food Chem.* **1975**, *23*, 495-498.
3. Buttery, R.G.; Ling, L.C.; Teranishi, R.; Mon, T.T. *J. Acric. Food Chem.* **1977**, *25*, 1227-1229.
4. Ha, J.K.; Lindsay, R.C. *Lebensm. Wiss. Technol.* **1990**, *23*, 433.
5. Ha, J.K.; Lindsay, R.C. *J. Food Sci.* **1991**, *56*, 1197-1202.
6. Sutherland, M.M.; Ames, J.A. *J. Sci Food Agric.* **1995**, *69*, 403-413.
7. Elmore, J.S.; Mottram, D.S.; Enser, M.; Wood, J.D. *Meat Sci.* **2000**, *55*, 149-159.
8. Schieberle, P. In: Characterization of food: emerging methods. (Goankar A.G., ed.) Elsevier Science BV, 1995, pp. 403-431.
9. Braggins, T.J. *J. Agric. Food Chem.* **1996**, *44*, 2352-2360.
10. Rota, V.; Schieberle, P. *J. Agric. Food Chem.* **2004**, submitted.

11. Schieberle, P. In: Flavour Precursors. Thermal and Enzymatic Conversions (Teranishi R., Takeoka G.R., Güntert M., eds.) ACS Symposium Series 490, Washington, DC, 1992, pp. 164-174.
12. Cerny, C.; Grosch, W. *Z. Lebensm. Unters. Forsch.* **1992**, *194*, 322-325.
13. Kerscher, R.; Grosch, W. In: Frontiers of Flavour Science (Schieberle P., Engel K.-H., eds.) Deutsche Forschungsanstalt für Lebensmittelchemie, Garching, Germany, ISBN 3-00-005556-8, 2000, pp. 17-20.

Texture

Chapter 7

Differential Retention of Emulsion Components in the Mouth after Swallowing: ATR FTIR Measurements of Oral Coatings

Harmen de Jongh[1,2], Anke Janssen[1,3], and Hugo Weenen[1,2,4]

[1]Wageningen Centre for Food Sciences, Diedenweg 20, 6700AN Wageningen, The Netherlands
[2]TNO Nutrition and Food Research, Utrechtseweg 48, 3700AJ Zeist, The Netherlands
[3]Agrotechnology and Food Innovations BV, Bornsesteeg 59, 6708PD, Wageningen, The Netherlands
[4]Current address: Numico R&D, P.O. Box 75338, 1118 ZN Schiphol Airport, The Netherlands

ATR FT-IR spectroscopy was used to investigate the coating that stays behind in the mouth after swallowing mayonnaise samples. Typically, the oil content was found to decrease to an average of 26% (range: 12-40%) of the t=0 value after 2 min, to <13% after 4 min and had practically disappeared from the oral cavity after 10 min. Unexpectedly, the carbohydrate/protein fraction of the oral coatings disappeared somewhat slower than the lipid fraction. For the xanthan-based sample, carbohydrate and protein were retained very well by the middle part of the tongue, but decreased quickly on the other parts of the mouth. For the starch-based product, carbohydrate and protein were retained well by the cheek and palate, but disappeared quickly from the tongue.

Creaminess and fatty afterfeel are amongst the sensory attributes that are most typical for emulsified foods (*1-3*). A recent study on texture characteristics of semi-solid foods (*3*) indicated that texture attributes of semi-solid foods can be grouped into six categories of 1. viscosity related attributes; 2. surface-feel attributes; 3. attributes related to bulk homogeneity/heterogeneity; 4. attributes related to ad/cohesion; 5. attributes related to sensations of wetness and dryness; and 6. attributes associated with oil sensations. The latter group of attributes, which is particularly related to the sensory functionality of oil, includes fatty, creamy and coating.

The textural characteristics of creaminess of oil-in-water food emulsion systems have been studied by several authors (*4-10*). Sensory studies and multivariate analysis (partial least square regression analysis) of a large series of model mayonnaises indicated that the sensory attribute 'creamy mouthfeel' positively correlates with the texture attributes 'fatty', 'thick' and 'sticky', and negatively with 'heterogeneous' (*11-12*). The texture attribute 'fatty' is particularly important for 'creamy mouthfeel', and can be differentiated in a pre-swallow (mouthfeel) and post-swallow component (afterfeel). The post-swallow component can be studied more easily, as it should be possible to isolate and analyze the stimulus responsible for the 'fatty afterfeel' and/or 'coating afterfeel' sensation present on the oral mucosa, i.e. the oral coating.

De Wijk *et al.* (*13*) have reported a methodology to study oral coatings by measuring the turbidity of oral water rinses. In this manner, they were able to correlate the turbidity values of oral rinses with the sensory attributes creamy, fatty, sticky and airy, for a series of dairy desserts varying in fat content between 0 and 15%. Prinz *et al.* (*14*) investigated directly (*in vivo*) the coating on the tongue using IR reflectance. Thus, they found good correlations between *in vivo* IR reflectance of food derived oral coatings on the middle $1/3^{rd}$ of the tongue and the texture attributes creamy, fatty and airy, also for a series of dairy desserts varying in fat content between 0 and 15% (*14*). However, these studies did not address the composition or compositional changes of these oral coatings. The study described here was intended to develop a method to quantitatively characterize the composition of oral coatings.

Materials and methods

Two mayonnaise samples containing 40% oil were used, one based on starch and one on xanthan gum. Composition was as follows: 40% soy oil, 5.6% liquid eggyolk, 3% of a 10% acetic acid solution (10%), 2% sugar, 0.3% mustard powder, 0.1% potassium sorbate, 87 ppm disodium EDTA and either 4% Farinex VA85T starch (AVEBE) or 1.3% xanthan gum; the rest was water. An emulsion was prepared using 2% egg yolk with a Koruma colloid mill (Disho

V 100/45) at 2800 rpm. This emulsion was gently mixed (Hobart mixer) with the water phase containing the starch or xanthan, the rest of the egg yolk and other ingredients.

Ten grams of a mayonnaise sample were circulated by tongue movements four times between tongue and palate and subsequently spat out (processing time of about 5 seconds). Next, either the mouth was rinsed with 25 mL of water, or swabs were taken from the tongue (middle and front part), cheeks and palate, using commonly available cotton wool covered sticks. The sticks were sealed to prevent evaporation. This experiment was repeated at least two times to check the reproducibility.

Samples were transferred from the swab sticks onto a germanium crystal (1x8 cm) and distributed evenly over the available area. The crystal was then placed in the light beam such that 6 total reflections were obtained. The sampling at six different spots on the crystal ensured a high reproducibility of the total signal. Moreover, when trimethylsilane was added to the dressing in test samples for intensity calibration purposes to judge the reproducibility of the material transfer it appeared that this varied by maximally 5% (results not shown). Attenuated total reflection infrared (ATR-IR) spectra were recorded on a nitrogen gas-flushed Biorad FTS 6000 equipped with a deuterated triglycine sulfate (DTGS) detector in a conditioned room at 20 °C (±0.2). Spectra were accumulated in the 4000-400 cm^{-1} region with a spectral resolution of 2 cm^{-1}, using a speed of 5 kHz and a filter of 1.2 kHz. The interferrograms were zero-filled prior to Fourier transformation to double the spectral resolution. The spectra were recorded relative to a clean crystal as reference. Typically 100 spectra were accumulated and subsequently averaged.

Every sample spectrum was first corrected for atmospheric water contributions. Next, the water contribution was subtracted by minimizing the net intensity of the water bending vibrations in the 2800-3400 cm^{-1} spectral region using an ATR-spectrum of pure water. Finally the spectra were smoothed using an apodization function of 5 data points.

Between samples the crystal was thoroughly cleaned using a detergent and distilled water. Sunflower oil, xanthan and albumin were chosen as representatives for three predominant constituents of the mayonnaises. All samples were prepared and measured at least twice.

Results and Discussion

Development of the method

Figure 1 shows the infrared spectrum of a 40% oil mayonnaise and the three reference compounds that represent the major constituents of the mayonnaise (except for water). Various vibration bands can be observed in the spectrum. The origins of the most characteristic bands are presented in Table 1. The spectral region between 3000 and 3600 cm^{-1} cannot be interpreted reliably due to the inadequate subtraction of the relatively large water contribution in this spectral range. The spectral 'gap' between 2500 and 2000 cm^{-1} is due to the presence of CO_2 in the atmosphere the concentration of which varies considerably in time.

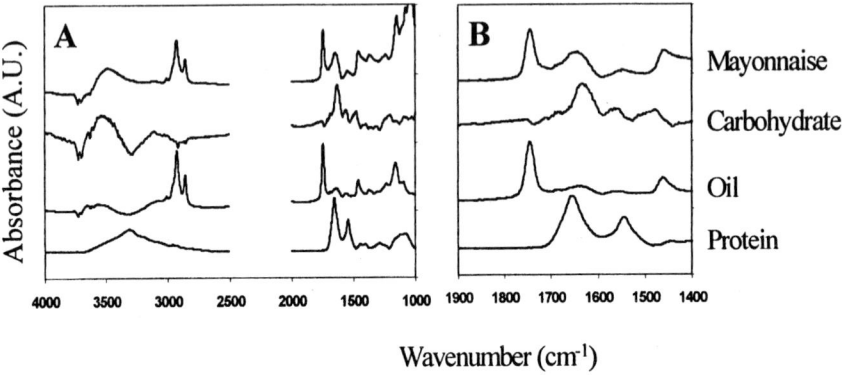

Figure 1. ATR-IR spectra of mayonnaise (top spectrum), a carbohydrate (xanthan; second spectrum), sunflower oil (third spectrum) and protein (egg albumin, bottom spectrum). Panel B is a selected region from panel A

Figure 1B shows that the spectrum of mayonnaise can be composed of contributions arising from an oil, a carbohydrate and a protein (and of course water that was already removed from the spectrum). Clearly, the intense oil vibrations (f.e. at 1750 and around 2800 cm^{-1}) are reasonably well resolved from the peaks of the carbohydrate and protein. Although the contribution of protein and carbohydrate can be distinguished in the complex mayonnaise, at this moment we limit ourselves in the analysis by taking those two signals together. It has to be stressed, however, that a quantitative refinement to resolve the ratio of protein to carbohydrate is possible, but requires careful baseline corrections and preferably an internal standard. To establish the quantity of oil versus carbohydrate/protein in mayonnaise samples, analysing the spectral region between 1900 and 1400 cm^{-1} is sufficient.

Table I. Typical infrared vibrations in mayonnaise

Frequency (cm^{-1})	Description	Compound
2950	CH$_3$ asymmetric stretch CH$_2$ anti-symmetric stretch	oil
2850	CH$_2$/CH$_3$ symmetric stretch	oil
1750	Ester carbonyl stretch	fatty ester
1660	Amide I carbonyl stretch	protein
1630	Hydroxyl-bending	carbohydrate
1550	Amide II N-H in-plane bending	protein

Figure 2A presents the difference of the spectrum of the first rinse after oral processing and that of the original (xanthan-based) mayonnaise. This difference spectrum is derived by weighting the two spectra such that the oil intensity around 1750 cm^{-1} becomes net zero. This subtraction procedure works quite accurately which is illustrated by the net zero intensity in the 3000-2700 cm^{-1} region, where the other oil contributions are present (not shown). The difference spectrum clearly shows that there is a net positive intensity in the 1700-1400 cm^{-1} region, indicating that the mouth-rinse has a different ingredient composition than the original sample, i.e. an enhanced concentration of protein and carbohydrate. Figure 2B shows an example of a physical sample taken from the cheek after oral processing of mayonnaise. Clearly in this swab the oil content is considerably higher than that of protein/carbohydrate compared to the original mayonnaise (see Figure 1). From the shape of the band between 1500 and 1700 cm^{-1} it can be concluded that there is no indication for a specific enrichment of either protein or carbohydrate in the sample.

It is evident from Figure 2 that the analysis of mouth rinses provides different results compared to those taken by swabs. Aqueous rinsing extracts more protein and carbohydrate compared to the oil phase. We conclude that the analysis of mouth swabs provides therefore a more realistic insight in the composition of oral coating compared to analysis of rinses.

The composition of the oral coating after mastication of mayonnaise samples

ATR FT-IR spectroscopy was used to investigate the oral coating that stays behind in the mouth after swallowing mayonnaise samples. To exemplify the method two mayonnaises were analyzed of which one was starch-based and the other xanthan-based, both containing the same amount of oil (40%). After oral processing, samples were taken from the front and middle part of the tongue, the cheek and the palate, after 2, 4, and 10 min. by taking swabs. The samples were then analyzed using ATR FT-IR spectroscopy, and absorptions representing oil, and protein /carbohydrate were quantified, as presented in Figure 3.

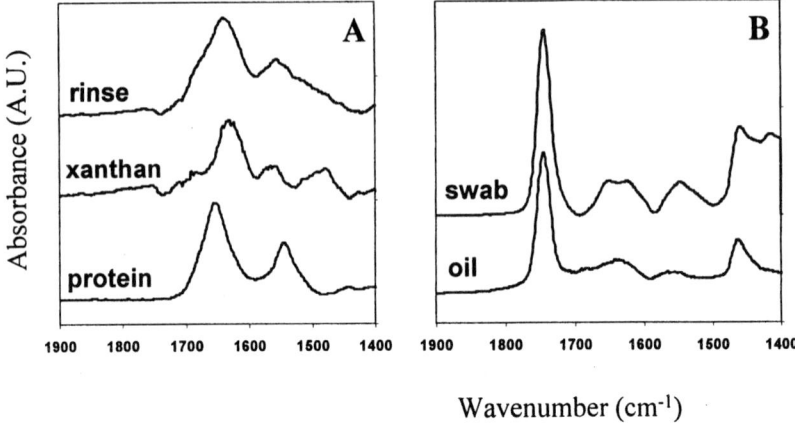

Figure 2. (Panel A) Difference spectrum of a mouth rinse after oral processing of a xanthan-based mayonnaise, together with reference spectra of xanthan and protein (Panel B) Example of a mouth swab (cheek; top spectrum) together with that of oil

For both mayonnaises it can be seen that the oil content was found to decrease to an average of 26% (range: 12-40%) of the t=0 value after 2 min., to <13% after 4 min. and had practically disappeared after 10 min. Moreover, there does not appear to be a significantly different retention of oil for the different parts of the mouth.

The oil content of the oral coatings appeared to decrease somewhat faster than the carbohydrate/protein fraction. This is not necessarily what was expected, since oil is less soluble in water, therefore less soluble in saliva, and would therefore not easily be cleared by dissolving in saliva to be swallowed. Apparently the situation is more complicated. There are of course other factors affecting clearance of the macronutrient components of an emulsion, like differences in adhesion affinity between the various constituents of the mayonnaise to the oral mucosa. It could also be postulated that the oil in mayonnaises is not 'free', but captured in emulsion-droplets that may be

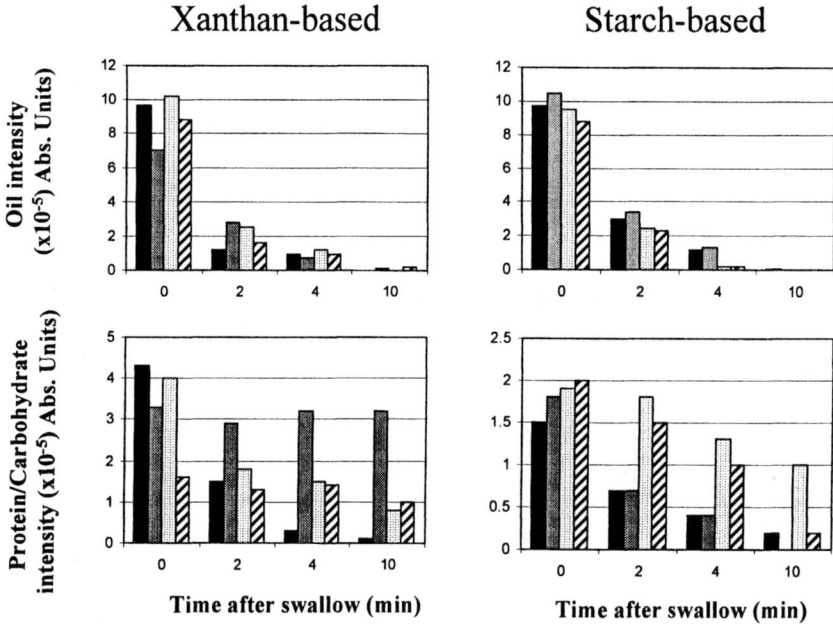

Figure 3. Intensity of oil (top panels) or protein/carbohydrate (bottom panels) of residual material on the tongue front (black bars), tongue middle (dark gray bars), cheek (light gray bars) and palette (dashed bars) as a function of time after swallowing a xanthan-based (left panels) and a starch-based (right panels) mayonnaise as analyzed by ATR FT-IR

stabilised by a protein coating. When, upon oral processing, this coating remains intact obviously the 'solubility' in saliva of the oil droplets would be greatly enhanced. Interestingly, the carbohydrate/protein content decreased differently when comparing the two thickeners used in this study (xanthan and modified starch), and for the different locations in the mouth. For the xanthan-based sample, carbohydrate and protein appeared to be retained well by the middle part of the tongue, especially in view of the second timescale relevant for human consumption. The front of the tongue appears to be cleared from mayonnaise constituents on a much faster timescale. For the starch-based product on the other hand the carbohydrate and protein were retained well by the cheek and palate, but disappeared rapidly from the tongue. The longer retention of starch (and protein) on the cheek and palette surface could point to the inability of amylase from the saliva to degrade these components as effective from these locations in the mouth compared to the tongue area. The roughness of the middle tongue surface might explain the different retention behavior of xanthan and starch, where the latter might be more able to penetrate into the cavities present on the tongue.

Acknowledgement

We thank Wilbert Oostrom for his technical assistance, and Jon Prinz and Lina Engelen for helpful suggestions when preparing for these experiments.

References

1. De Wijk, R.A.; Engelen, L.; Prinz, J.F. *Appetite* **2003**, *40*, 1-7.
2. Kilcast, D. and Clegg, S. *Food Qual. Pref.* **2002**, *13*, 609-623.
3. Weenen, H.; van Gemert, L.J.; van Doorn, J.M.; Dijksterhuis, G.B.; de Wijk, R.A. *J. Texture Studies* **2003**, *34* (2), 159-179.
4. Clegg, S.; Kilcast, D.; Arazi, S. In *Proceedings of third International Symposium on Food Rheology and Structure*, 2003, 373-377.
5. Daget, N.; Joerg, M.; Bourne, M. *J. Texture Studies* **1987**, 18: 367-388.
6. Daget, N.; Joerg, M. *J. Texture Studies* **1991**, *22*, 169-189.
7. Kokini, J.L. *J. Food Engr.* **1987**, *6*, 51-81.
8. Wendin, K.; Aaby, K.; Edris, A.; Risberg Ellekjaer, M.; Albin, R.; Bergenstahl, B.; Johansson, L.; Pilman Willers, E.; Solheim, R. *Food Hydrocolloids* **1997**, *11*, 87-99.
9. Stern, P.; Valentova, H.; Pokorny, J. *Eur. J. Lipid Sci. Technol.* **2001**, *103*, 23-28.
10. Weenen, H.; Jellema, R.H.; de Wijk, R.A. In *Flavour and texture of lipid containing foods;* Weenen and Shahidi, Eds.; Am. Chem. Soc.: Washington DC, 2005.
11. De Wijk, R.A.; Van Gemert, L.J.; Terpstra, M.E.J.; Wilkinson, C.L. *Food Qual. Pref.* **2003**, *14*, 305-317.
12. Weenen, H.; Jellema, R.H.; de Wijk, R.A. *Food Qual. Pref.* **2005**, *16*, 163-170.
13. De Wijk, R.A.; Prinz, J.F.; Janssen, A.M. *Food Qual. Pref.* **2005**. Submitted.
14. Prinz, J.F.; de Wijk, R.A. **2004**. Sensory correlates of lingual coatings following the ingestion of semi-solid foods. Unpublished results.

Chapter 8

The Role of Fats in Friction and Lubrication

J. F. Prinz[1,2], R. A. de Wijk[1,3], and H. Weenen[1,2,4]

[1]Wageningen Centre for Food Sciences, P.O. Box 557, 6700 AN Wageningen, The Netherlands
[2]TNO Nutrition and Food Research, P.O. Box 360, 3700 AJ Zeist, The Netherlands
[3]A&F, P.O. Box 17, 6700 AA Wageningen, The Netherlands
[4]Current address: Numico R&D, P.O. Box 75338, 1118 ZN Schiphol Airport, The Netherlands

The sensory space for semi-solid foods is dominated by two dimensions, one running from thick to melting, the other from creamy and fatty at one end to a cluster of attributes comprising astringency, dry-mealy, smooth-grainy, prickling and roughness. Friction seems to be the major determinant of creaminess - rough axis. In the studies described here we demonstrate a logarithmic relationship of fat content with friction (r^2=0.9) using a wide range of semi-solid foods with fat content of 0 to 82%. Coefficients of kinetic friction ranged from 1.2 for the zero fat product, to 0.4 for the highest fat product. We also demonstrated an effect of fat droplet size on friction using a series of model mayonnaises which had identical ingredients but were processed to afford a range of droplet sizes. Coefficients of friction varied from 0.69 to 0.75 and were positively correlated with droplet size.

Friction is the force that resists the movement of one surface against another. Two forms of friction are recognized, static friction – the force required to initiate movement and kinetic friction - the force required to keep an object in motion, kinetic friction is lower than static friction. The coefficient of kinetic friction is defined as the ratio of the vertical force applied to the surface to the horizontal force required to maintain motion. Friction is a system property, not a material property, depending on the physical properties of the two surfaces, their surface profile and the properties of any interposed material which may act as a lubricant.

There are two "laws" of friction; Amonton's law states that friction is independent of the macroscopic contact area; Coulomb's law states that friction is independent of speed. Neither law holds for all combinations of materials under all conditions (*1*). It is therefore important to measure friction under realistic conditions and to use appropriate materials.

Frictional conditions in the mouth have been implicated in perception of important food attributes such as astringency, mealiness, smoothness, roughness and slipperiness, among others (*2-4*). This ability to detect friction probably evolved so that foods which could wear the teeth excessively could be avoided (*5*) and to serve as a mechanism by which fats and oil could be detected. In humans, sensations related to friction and lubrication affect consumer responses to food products (*2,3*) and are therefore commercially important. Despite the generally accepted recognition of the role of friction in determining the sensory performance of food product there have been very few attempts to quantify the tribological properties of foods.

Ollsen *et al.* (*6*) developed a method for measuring the coefficient of friction of oral mucosa *in vivo*, as a means of assessing the efficacy of artificial saliva, and to provide objective criteria for the evaluation of xerostomia (dry mouth). This device consisted of a rotating stainless steel disc that was held against the mucosa of the everted lower lip. The force, applied perpendicular to the oral surface and the torque required to rotate the disc were continuously monitored, thereby allowing the frictional coefficient to be computed. Values obtained ranged from a maximum of 14 in a totally xerostomic subject to a mean 0.4 in subjects with normal flow rates (*7*). Although these results clearly demonstrate the importance of saliva in lubricating the mouth they do not assess the frictional forces experienced by the oral mucosa during normal function, since the behaviour of mucosa vs. mucosa is not necessarily the same as steel vs. mucosa, to the extent that the lubricity of different lubricants may not have the same rank order under different conditions.

The mechanisms that generate frictional forces are not fully understood, and it should always be remembered that friction is a system property, not a material property, so it is inappropriate to say that rubber for instance has high friction and Teflon has low friction, rather one must specify the two surfaces involved, their surface roughness, their relative speed, the loading conditions and the properties of any intervening material which may act as a lubricant. Therefore if one wishes to test the efficacy of a lubricant it must be performed under conditions relevant to the system of interest.

Lubrication by saliva

The apparatus shown in figure 1 was constructed to measure lubrication, *in vitro*. The device is modified from Halling (*8*) it consisted of a rubber band (6 cm length x 1 mm^2) attached to a load cell. One end of the rubber band was looped round the metal cylinder of an electric motor; the other was attached to the load cell (see Figure 1). When the motor was switched on and rotated clockwise, friction between the cylinder and the rubber band produced a load, F_1, that could be detected at the load cell. When the direction of the cylinder was reversed, the load then dropped to F_2. If loads F_1 and F_2 are both known, then it can be shown (*8*) that the coefficient of friction (μ) is given by the formula $\mu = 1/\pi \ln (F_1/ F_2)$. To make a measurement the surface of the cylinder was covered by a layer of the sample and the tension in the rubber band was recorded as the cylinder was rotated clockwise and anticlockwise. All measurements were made at 20 °C.

The rubber band was refreshed after each measurement. Three replicates were made for each sample. Friction was measured on foods as is, with 15% by vol. saliva added, or with the same amount of water added. At least 1 hour was allowed for amylase in the saliva to completely breakdown any starch..

Experiment 1

Stimuli. Five white sauces, five mayonnaises, and five vanilla custard desserts were used with fat contents ranging from 0 to 72% (see Table I). All products were commercially available. White sauces were prepared according to the instructions but variation in fat content was increased by using either water or full fat milk plus additional butter for the preparation of two of the Knorr white sauces. Mayonnaises and custards were refrigerated at 6°C prior to consumption, whereas the sauces were heated to 60°C by micro waving. All foods were served to the subjects in white polystyrene cups. The subjects used disposable dinner spoons to sample the foods.

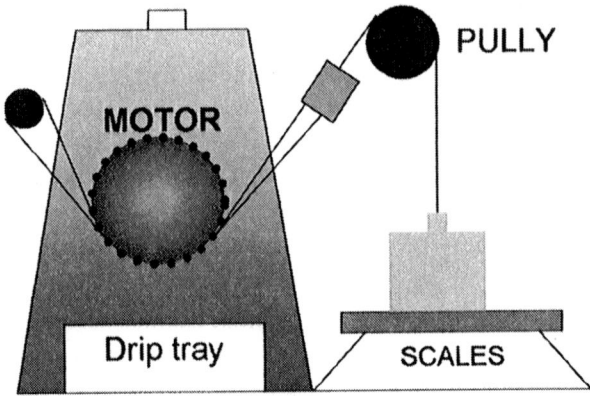

Fig 1. Schematic representation of apparatus used to measure lubrication

Table I. Semi-solid foods used in study 1

Food-type	% Fat	Brand
Custard	1.8	Alpro soya-dessert vanille
Custard	0.5	Creamex Magere vanillevla (UHT)
Custard	2.6	Friesche Vlag Vanillevla
Custard	3.6	Friesche Vlag Romige vanillevla
Custard	3.0	AH Vanillevla
Mayo	36.0	C1000 Mayohalf
Mayo	52.0	D&L Mayonaise with lemon
Mayo	72.0	Calvé Mayonaise
Mayo	16.0	Remia Friteslijn
Mayo	40.0	Becel Dressing
Sauce	0.0	Knorr white sauce with water
Sauce	1.8	Grand Italia Besciamellasaus
Sauce	24.8	Knorr white sauce with full milk and butter
Sauce	4.4	Knorr cream sauce
Sauce	8.5	Knorr Hollandaise sauce

Panelists / sensory attributes. The sensory properties of the products were investigated with the use of a sensory panel trained according to the principles of Quantitative Descriptive Analysis. Panelists were selected from a group of 200 young and healthy candidates. Selection tests included an odor identification test, an odor memory test, a verbal creativity test, and a series of texture tests in which panelists' abilities to assess fattiness, roughness, and particle size were measured. Seven panelists aged between 30 and 48 years and with above-average scores on all tests were selected for the trained panel. The panelists were paid for their participation and testing took place at the sensory facilities of TNO Nutrition and Food Research, Zeist, the Netherlands. Panelists were seated in sensory booths with appropriate ventilation and lighting. During nine previous 2-h sessions, the panel had been trained with commercially available vanilla custard desserts, white sauces, and mayonnaises to establish a list of 35 appropriate odor, flavor, mouthfeel, and afterfeel attributes (2).

Procedure: The panelists produced sensory profiles for the test foods, presented once in a random order at a rate of one food per 5 min, each food was presented in three replicates. The food was first smelled after which odor

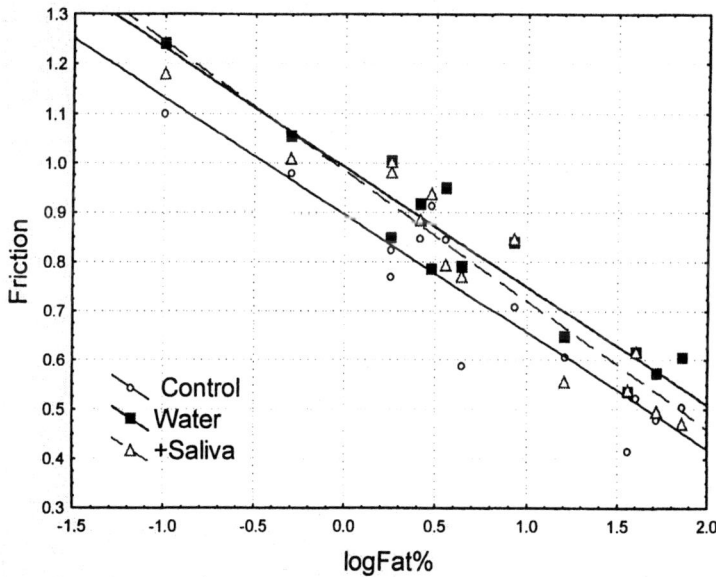

Figure 2. Relationship of friction to % fat for custards, mayonnaises and sauces.

attributes were rated. Next, the product was taken into the mouth and taste/flavor and mouthfeel attributes were rated in the order in which they were perceived. Finally, the product was swallowed and afterfeel attributes were rated. Acquisition of panelists' responses was done by computer using FIZZ software (Biosystemes 1998, v1.20K, Couternon, France). The attributes appeared by category on a monitor in front of the panelists which listed attributes on the left, and on the right a 100-point response line scale anchored at the extremes. Panelists used a mouse to indicate the perceived strength of each attribute.

Results

There was a highly significant correlation of the coefficients of friction with log % fat ($r^2 = 0.872$, 0.91 and 0.91 for the control, water and saliva conditions; see Figure 2). Coeficients of kinetic friction were significantly lower for the unmodified controls in comparison to the other two conditions, whereas there was no significant difference between the samples to which water and saliva had been added. This suggests that in these products starch breakdown by salivary

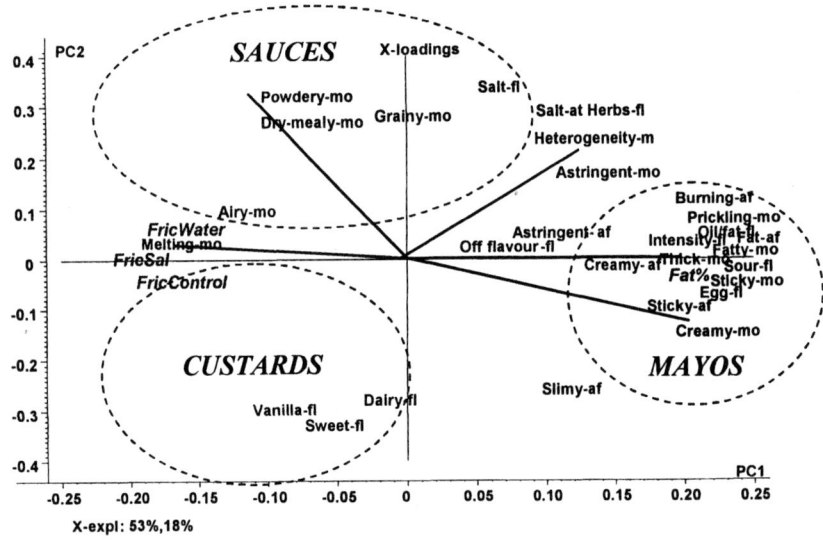

Figure 3. PCA Biplot showing the inter-relationship of the sensory attributes with friction and fat content for the product set shown in Table I.

amylases did not play a significant role in affecting the friction. This is contrary to previous findings in our laboratory where starch based model custards were investigated. With these products there was a dramatic reduction in the products viscosity, and a correspondingly large change in friction. We attribute the lack of an effect of saliva induced breakdown to the presence of thickeners other than starch in these products. Friction was significantly correlated with melting. The inter-relationship of the instrumental friction and sensory reatings are shown in the PCA bi-plot shown in Figure 3.

Experiment 2

In a second experiment a series of model mayonnaises were investigated; these had identical ingredients, but processing was modified in order to control oil droplet size. Specifically, to produce emulsions with small droplets, all egg yolk was added prior to processing, whilst to produce large droplets a small amount of egg yolk was added before processing and the remainder was slowly mixed in later. Fat droplet sizes were measured using light microscopy (Figure 4). In this series coefficients of kinetic friction ranged from μ= 0.69 to 0.75. There was a highly significant correlation of droplet size with friction. (Figure 5).

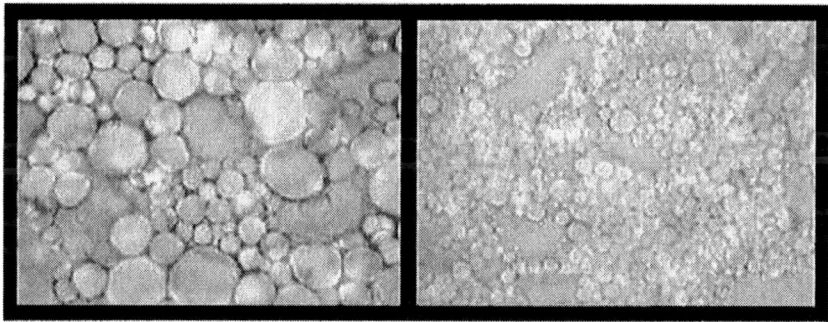

Figure 4. Model mayonnaises made with identical ingredients, (40% oil) but processed to give different oil droplet sizes.

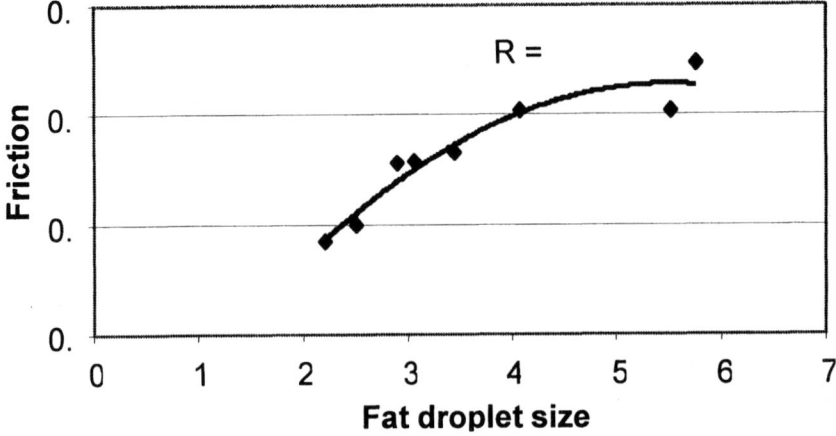

Figure 5. Correlation of fat droplet size with friction in a series of 40% fat model mayonnaises with varying droplet sizes

Conclusions

The measurement system used here attempts to mimic conditions in the mouth by using two contrasting surfaces. The rubber band provides an analog for the relatively rough and compliant surface of the tongue, the aluminum shaft approximates to the smooth less deformable palatal mucosa (9). Although the system is far from a faithful representation of the oral conditions, the system appears to give results which correlate well with sensory findings. The results from the first experiment clearly show the capacity of fats and oils to act as lubricants. However lipids are not the only materials capable of acting as lubricants, carboxmethyl cellulose gels for example are also very good lubricants (μ=0.1). Furthermore, as demonstrated in the second experiment, the degree of lubrication provided by fat droplets can be significantly manipulated by controlling their diameter. By taking friction and lubrication into account when designing new products, food technologists may be able to provide the desirable sensory properties of high fat products whilst reducing the actual fat content.

References

1. Comaish, S.; Bottoms, E. *Br. J. Derm.* **1971**. *84*, 37-43.
2. De Wijk, R.A.; Prinz, J.F. *Food Qual. Pref.* **2005**, *16(2)*, 121-128.
3. Green, B.G. *Acta Psych.* **1993**, *84*, 119-25.
4. Kokini J.L. *J. Food Eng.* **1987**, *6*, 51-81.
5. Prinz J.F. Abrasives in foods and their effect on intra-oral processing: A two-colour chewing gum study. *J. Oral. Rehab.* (In press)
6. Olsson, H.; Henricsson, V.; Axell, T. *Scand. J. Dent. Res.* **1991**, *9*, 329-32.
7. Olsson, H.; Spak, C.J.; Axell, T. *Acta Odontol. Scand.* **1991**, *49*, 273-9.
8. Halling, J. **1976**. Introduction to tribology. London: Wykeham.
9. Sivamani, R.K.; Goodman, J.; Bitis, N.V.; Maibach, H.I. *Skin Res. Techol.* **2003**, *9*, 235-239.

Chapter 9

Prediction of Creamy Mouthfeel Based on Texture Attribute Ratings of Dairy Desserts

H. Weenen[1,2,4], R. H. Jellema[2], and R. A. de Wijk[1,3]

[1]Wageningen Centre for Food Sciences, P.O. Box 557, 6700 AN Wageningen, The Netherlands
[2]TNO Nutrition and Food Research, P.O. Box 360, 3700 AJ Zeist, The Netherlands
[3]A&F, P.O. Box 17, 6700 AA Wageningen, The Netherlands
[4]Current address: Numico R&D, P.O. Box 75338, 1118 ZN Schiphol Airport, The Netherlands

A quantitative predictive model for creamy mouthfeel in dairy desserts was developed, using PLS multivariate analysis of texture attributes. Based on 40 experimental custard desserts, a good correlation was obtained between measured and predicted creamy mouthfeel ratings. The model was validated by testing it for commercial custard desserts ($r=0.84$, $p=0.002$). Further validation was obtained by applying the model to commercial yoghurts, using a different panel. Again a good correlation was obtained ($r=0.90$, $p<0.001$) after re-calibrating the model, which was necessary, as the panels were not trained to apply the same scaling behavior. Texture attributes which were most important in the PLS model for creamy mouthfeel included thick, airy and fatty mouthfeel with positive coefficients, and rough and dry mouthfeel with negative coefficients.

Texture characteristics of semi-solid foods (*1*) can be grouped into six categories: 1. viscosity related attributes; 2. surface feel attributes; 3. attributes related to bulk homogeneity/heterogeneity; 4. attributes related to ad/cohesion; 5. attributes related to sensations of wetness and dryness; and 6. attributes associated with fat sensations. The latter group of attributes are particularly related to the sensory functionality of fat, and include fatty, creamy and coating. Creamy is a particularly interesting attribute, as it is generally well correlated with consumer preference (*2-6*) and because it is a complex or composite attribute.

Several studies have addressed the nature of creaminess, both in terms of the underlying sensory sub-attributes, and in terms of the underlying rheological and mechanical properties. Prentice (*7*) reported that granularity or grittiness is a negative factor. A study by Wood (*8*) suggested that creamy textured soups should have a very smooth mouthfeel, with complete absence of powderiness.

In a study on mouthfeel characteristics of beverages, Szczesniak (*9*) classified creamy in a category called feel on soft tissue surfaces, together with smooth and pulpy. Studies with mainly dairy products (*10*) indicated that creaminess can be predicted based on only two other attributes, i.e. thick and smooth ($r=0.90$), according to the following formula: log creamy = 0.54log thick + 0.84log smooth. Clearly thick and smooth are important factors, however the accuracy of the creamy prediction using this formula seems to depend on the choice of products, as a previous study with other products (*11*) had resulted in a much lower r (0.68). This led the authors to conclude that creamy is more than a combination of smooth and thick.

De Wijk *et al.* (*12*) reported on creaminess in custard desserts varying in fat content, carrageenan and starch. Partial least squares analysis of creamy mouthfeel based on sensory attribute ratings showed positive regression values for creamy flavor, fatty flavor, thick mouthfeel, fatty mouthfeel and fatty afterfeel, whereas bitter chemical flavor, sickly flavor and rough mouthfeel had negative regression coefficients. Two latent variables explained 98% of the variance based on these sensory attributes.

In the study described below, linear multivariate (Partial least squares) models were developed based on experimental custard desserts, and these models were tested and validated with commercial custard desserts and yoghurts.

Experimental

Samples investigated in this study were either prepared at NIZO Food Research, Ede, The Netherlands, or purchased from local supermarkets. Samples were served directly from the refrigerator in foam cups covered with a lid.

Plastic dessert spoons were available for eating. Full details of the composition and processing are given in Tables I-III. The composition of the experimental custard desserts were selected based on the following criteria: the Ve and Vf series were an experimental design, with starch type and content, carrageenan type and content, and fat level as variables. The Vh series was designed to study the effects of the type of starch. The Vi series was designed to vary the fat and starch level. The modified starches used in this study were of tapioca (VA85T), potato (VA20, VA40 & VA70) and maize (WM50) origin, and were obtained from Avebe (Farinex type; Wormerveer, The Netherlands). Processing: included no homogenization, only UHT heating for 5 seconds at 144°C. All experimental desserts were flavored with 0.1% vanillin powder, which was obtained from Danisco.

Assessors were selected for their sensory capabilities, trained with the samples that were later used for measurement and paid for their participation. Emphasis in the training was on mouthfeel and afterfeel attributes, odor and flavor attributes were kept limited. Details of the sensory attributes and procedure used has been described elsewhere (1). All custard desserts were evaluated by a descriptive panel (12 subjects) in tripicate, using attributes that were determined by that same panel (1). The Ve, Vf, Vh and Vi series were each evaluated in separate sets of each 3 sensory evaluation sessions.

Odor attributes included: odor intensity, sour odor, cream odor and off-dour. Mouthfeel attributes included: cold, thick, airy, grainy, creamy, sticky, heterogeneous, dry/mealy, melting, fatty, powdery, prickling and rough/astringent. Flavor attributes included: flavor intensity, sweet, salt, sour, creamy, sour milk, bitter, flour, off-flavor. Afterfeel attributes included: creamy, sticky, prickling, astringent/rough and mouth-watering. One aftertaste attribute was used: sour. ANOVA indicated significant product effects ($p < 0.05$) for all these attributes except for salty and off-odor, which were not significant in all product categories.

The yoghurts were evaluated by another descriptive panel. The yoghurt panel (12 subjects) was trained with the attributes and attribute definitions that were generated in the custard dessert study but developed two additional attributes, buttermilk favour, and musty/stale flavor.

Multivariate data analysis was performed in Matlab 6.1.0.450 (The Mathworks, Inc.) using the PLS Toolbox 2.0.1b (Eigenvector Research Inc.). Partial least squares (PLS, see refs. *13-14*) modelling of creamy mouthfeel as a function of texture mouthfeel and afterfeel attributes was used to obtain the regression factors for creamy mouthfeel. PLS is a method to perform multivariate calibration. The PLS latent variables are linear combinations of the original variables. The criterion to select weights for the linear combination is that the latent variables describing the x-data (mean texture attribute ratings) should have maximal covariance with the y-vector (creamy mouthfeel in this

Table I. Composition of experimental desserts

Code	Starch	%	Carrageenan type	%	Fat in milk (%)
Ve01	VA85T	3.8	-	-	0
Ve02	VA85T	3.8	K	0.02	4.5
Ve03	VA40	4.5	-	-	4.5
Ve04	VA40	4.5	L	0.025	0
Ve05	VA85T	4.5	-	-	4.5
Ve06	VA85T	4.5	K	0.02	0
Ve07	VA40	3.8	-	-	0
Ve08	VA40	3.8	L	0.025	4.5
Vf01	VA40	3.8	-	-	4.5
Vf02	VA40	3.8	K	0.02	0
Vf03	VA85T	4.5	-	-	0
Vf04	VA85T	4.5	L	0.025	4.5
Vf05	VA85T	3.8	-	-	4.5
Vf06	VA85T	3.8	L	0.025	0
Vf07	VA40	4.5	-	-	0
Vf08	VA40	4.5	K	0.02	4.5
Vf09	VA85T	4.5	-	-	4.5
Vh01	VA85T	4.5	-	-	3
Vh03	VA20	3.8	-	-	3
Vh05	VA70	4.3	-	-	3
Vh07	VA20	3.6	-	-	3
Vh08	VA50T	3.3	-	-	3
Vh09	VA70	4.5	-	-	3
Vh12	WM50	4.0	-	-	3
Vi03	VA85T	4.5	-	-	0
Vi04	VA85T	4.7	-	-	0
Vi01	VA85T	4.9	-	-	0
Vi20	VA85T	5.1	-	-	0
Vi10	VA85T	4.3	-	-	5
Vi11	VA85T	4.5	-	-	5
Vi12	VA85T	4.7	-	-	5
Vi13	VA85T	4.9	-	-	5
Vi23	VA85T	4.1	-	-	10
Vi24	VA85T	4.3	-	-	10
Vi25	VA85T	4.5	-	-	10
Vi26	VA85T	4.7	-	-	10
Vi17	VA85T	3.9	-	-	15
Vi18	VA85T	4.1	-	-	15
Vi19	VA85T	4.3	-	-	15
Vi21	VA85T	4.5	-	-	15

Table II. Information on the composition of the commercial custard desserts

Code	Product name	% Fat	Thickeners	Milk source
C1	Coberco romige vla	3.6	Maize starch, mod. maize starch, carrageenan	Dairy
C2	AH vanille vla	3	Starch, mod. starch, carrageenan	Dairy
C3	Melkunie boeren vla	3	Modified tapioca starch, carageenan	Dairy
C4	Melkunie vanille vla	3	Maize starch, mod. maize starch, carrageenan	Dairy
C5	Zuivere zuivel vanille vla	3	Maize starch	Dairy
C6	AH Blanke vla	3	Maize starch, mod. maize starch, carrageenan	Dairy
C7	Coberco vanille vla	2.6	Maize starch, mod. maize starch, carrageenan	Dairy
C8	AH Biologische vla	2.5	Maize starch, carrageenan	Dairy
C9	Alpro soja vla	1.8	Mod. starch, carageenan	Soy
C10	Euroshopper	<0.5	Starch, mod. starch, carrageenan	Dairy

Table III. Information on the composition of the commercial yoghurts

Code	Product name	Fat (%)	Protein (%)	Carbohydrate (%)
Y1	Total full fat	10	6	4
Y2	Turkish full fat	10	3.7	3.5
Y3	Nestle Greece	9.2	4	4.7
Y4	AH bio stand	3.5	4	4.5
Y5	AH bio roer	3.5	4	4
Y6	Turkish low fat	3.5	4	3.7
Y7	AH full fat	3	3.5	3.5
Y8	AH full fat zachte	3	3.5	4
Y9	Campina Boerenland	3	5	5.5
Y10	Groene Koe	3	3.5	4
Y11	Vitamel LGG	3	5	5
Y12	Campina bio halfvol	1.5	4.5	4.5
Y13	Mona Bulgaarse	0.5	5.5	4.5
Y14	AH no fat	0	4	4
Y15	AH no fat zachte	0	4	4
Y16	Total no fat	0	10	4

case). The aim of PLS is to model y as a function of x. Just like in principal component analysis (PCA), linear combinations of the original variables are obtained but the difference is that optimal co-variance with the y-vector is obtained.

The scores averaged over replicates and panel members, were mean centred. The following terms are used throughout the text: x-block, meaning the matrix containing the scores of the descriptors given by the panel members except creamy mouthfeel; y-vector, meaning the vector representing the ratings of creamy mouthfeel by the panel members; latent variable, meaning the new coordinate system calculated by means of PLS describing the relationship between the variables in the x-block and the y-vector.

As the present study was aimed at understanding the textural requirements of creamy mouthfeel, models were developed and investigated for creamy mouthfeel versus other textural attributes.

When the custard dessert PLS model for creamy mouthfeel was applied to the commercial yoghurts, the prediction clearly deviated from the creamy mouthfeel ratings that were actually measured for the yoghurts. The reason for this is most likely that a different panel was used to obtain the yoghurt results. This panel was trained using the same attributes and definitions for attributes, however their scaling behavior was not trained, and was apparently different. In addition, yoghurts are more sour than custard desserts, which may affect creaminess and therefore the creamy model. The differences (residuals) between the model and the measured values correlated well with creamy mouthfeel, indicating that the deviation is due to a calibration difference. The custard model was therefore transformed, to obtain maximum fit. This resulted in the following transformation function: $y = 2.29 + 0.65x$ in which y is the predicted creaminess based on the custard dessert PLS model and x is measured creaminess of the commercial yoghurts.

Results and Discussion

When combining results of different product groups or of different measurements in one PLS model, it is important that these data cover the same multivariate space. If not, results may be obtained that are dominated by differences between the groups of measurements, and the predictions are extrapolations rather than interpolations within the space of the training set. To determine to what extent the products in this study are in the same multivariate sensory space, the score plot for the PLS models of creamy mouthfeel versus selected texture attributes was made (Figure 1). The resulting figure shows that

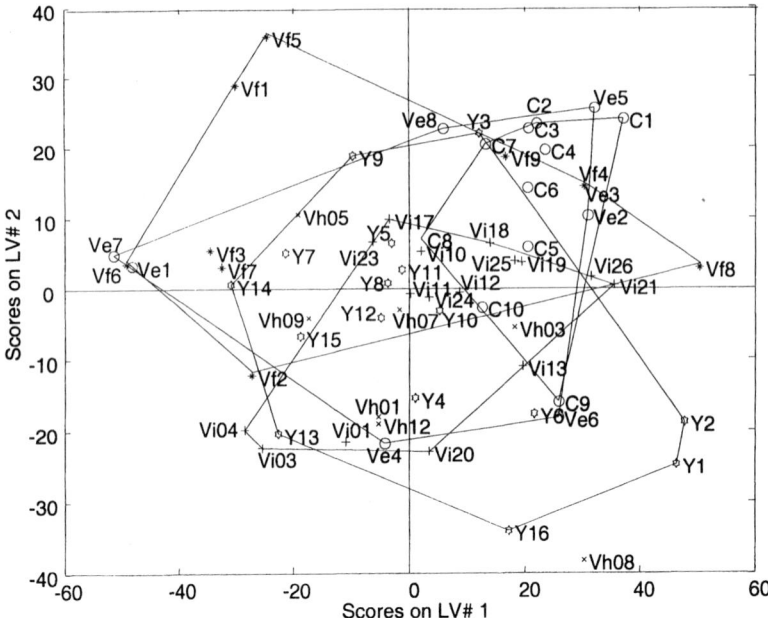

Figure 1. Score plot showing the resemblance of the two-dimensional sensory space of the experimental custard desserts (Ve, Vf, Vh, Vi), commercial custard desserts (C) and commercial yoghurts (Y) resulting from PLS regression of creamy mouthfeel versus selected texture attributes (LV = latent variable)

all commercial custard desserts and most commercial yoghurts are part of the sensory space of the experimental custard desserts, except two high fat yoghurts (Total full fat, Y1; Turkish full fat, Y2), and one zero fat yoghurt (Total no fat, Y16). Fat content as such can therefore not be a reason for the exceptional behavior of these 3 yoghurts. A possible explanation may be that the custard desserts contain starch, which breaks down in the mouth (*15*), and yoghurts normally do not contain starch. Another factor may be that sourness is an important attribute for yoghurts but not for custards. Nevertheless considering all the obvious differences between commercial yoghurts, commercial custard desserts and the experimental custard desserts, their sensory space was sufficiently homogeneous and overlapping to develop predictive sensory models.

Previously we have determined which texture attributes contribute to creamy mouthfeel, in a more qualitative way (*12, 16*). This study indicated that the following texture attributes consistently correlated with creamy mouthfeel in

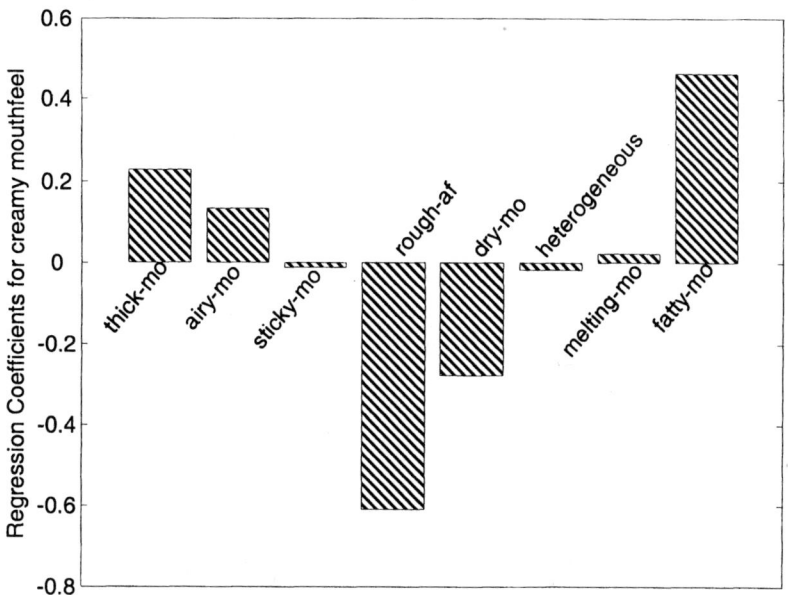

Figure 2. Loads plot for PLS regression of creamy mouthfeel versus selected texture attributes, showing the values of the regression coefficients

rough afterfeel, dry mouthfeel, heterogeneous mouthfeel, melting mouthfeel and fatty mouthfeel. These texture attributes were used in this study to develop a quantitative model for creamy mouthfeel, using Partial Least Squares analysis. Figure 2 shows that thick, airy and fatty are strongly positively correlated with creamy mouthfeel, and that rough afterfeel and dry mouthfeel were strongly negatively correlated with creamy mouthfeel. Using the regression coefficients resulting from the PLS regression of creamy mouthfeel vs selected texture attributes for the experimental custard desserts (Figure 2), a good correlation was found for the measured creamy mouthfeel ratings versus the predicted creamy mouthfeel ratings (Figure 3; $r= 0.94$, $p < 0.01$).

To test whether this model is valid in an independent data set, this same model was used for the prediction of creamy mouthfeel of commercial custard desserts. Again a good correlation was obtained (Figure 4; $r= 0.85$, $p= 0.002$), confirming the validity of the model for custard desserts in general. The model was also applied to commercial yoghurts, to test whether it can also be applied to other dairy product categories. For yoghurts a different panel was used,

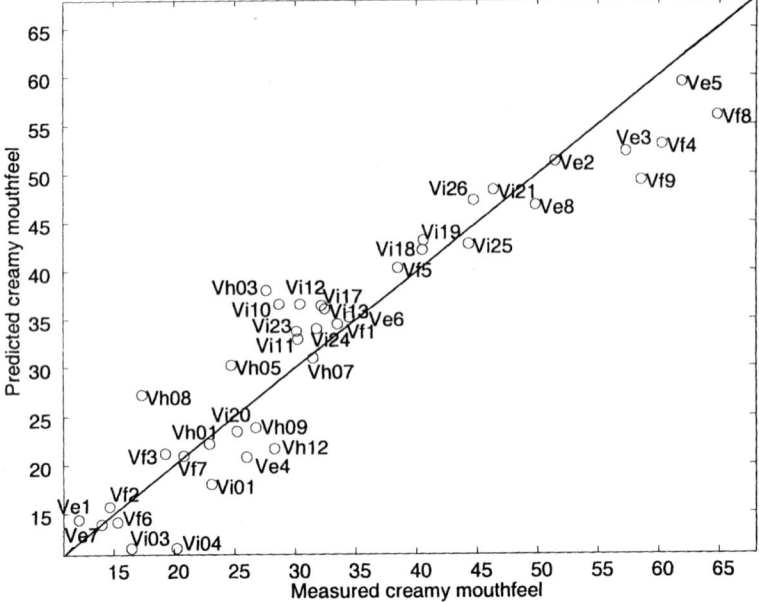

Figure 3. Scores plot for creamy mouthfeel model based on experimental custard desserts, predicted creamy mouthfeel values versus measured values

which was trained to use the same attributes, but which was not trained to apply the same scaling behavior as in the custard dessert study. The results therefore had to be re-calibrated, to make up for differences in scaling behavior (see Methods). The resulting model predicted creamy mouthfeel in the yoghurts very well (Figure 5; r= 0.90, p<0.001), indicating that this model for creamy mouthfeel is applicable to various (dairy) product categories.

The model for creamy mouthfeel described here was based on texture attributes only. However, creamy mouthfeel is strongly correlated with creamy flavor (*16*), and with other flavor attributes (*12*). Nevertheless good correlations were obtained for predicting creamy mouthfeel based on only texture attributes, versus measured creamy mouthfeel (Figure 3). Similar to the results obtained by de Wijk *et al*. (*12*), positive regression coefficients were obtained for thick and fatty mouthfeel. De Wijk *et al.* (*12*) found a negative regression coefficient for

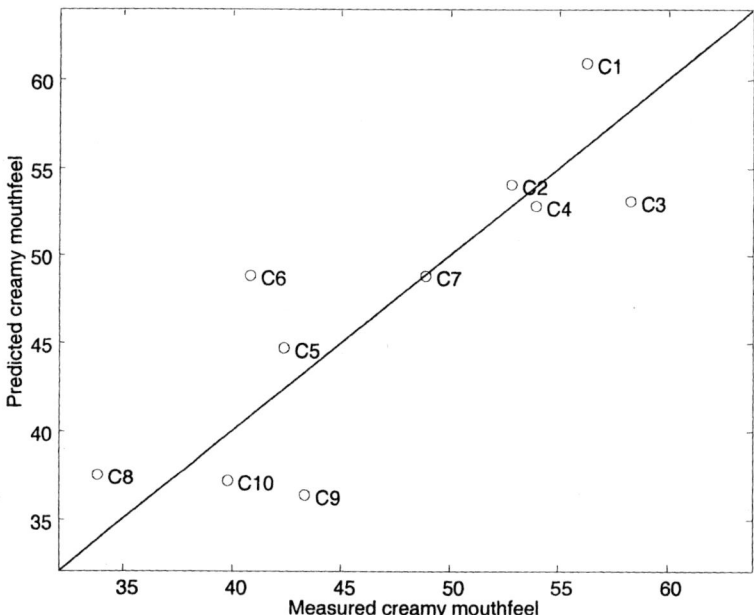

Figure 4. Scores plot for creaminess in commercial custard desserts predicted by model based on experimental custard desserts

rough mouthfeel, this study indicated a strong negative regression coefficient for rough afterfeel. Rough mouthfeel and rough afterfeel are generally well correlated (*1*), so it is to be expected that rough mouthfeel and rough afterfeel are to some extent exchangeable.

In addition, this study indicated a positive correlation for creamy mouthfeel with airy, and a negative correlation with dry. Unlike in previous sensory PLS studies on creamy mouthfeel (*16*), regression coefficients for sticky, heterogeneous and melting were negligible. This indicates that these attributes show little co-variation with creamy mouthfeel for the investigated products.

One of the most deviating commercial custard desserts in the prediction of creamy mouthfeel, was the only soy based custard dessert in this study (sample C9 in Figure 4). Interestingly, measured creaminess for this product was considerably higher than predicted creaminess. The reason for this is not clear.

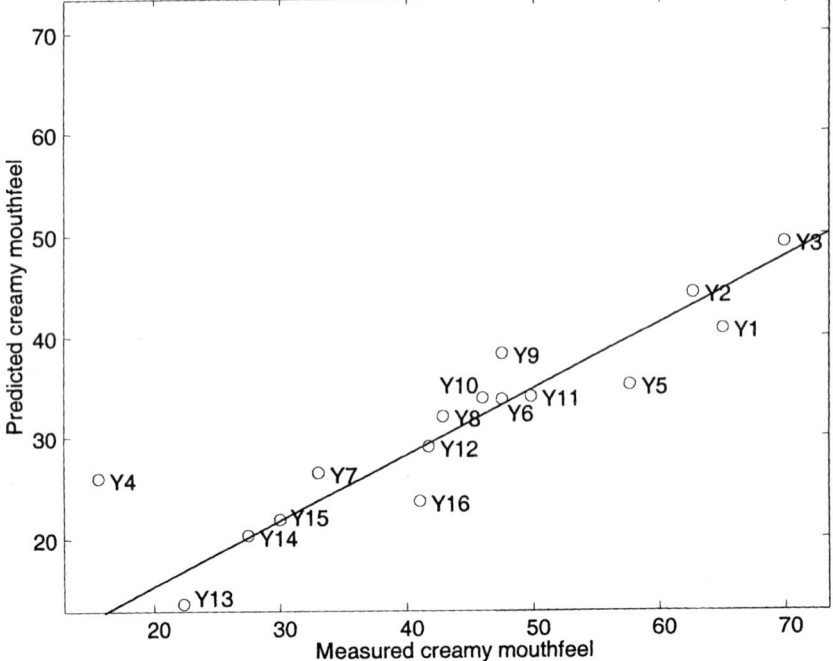

Figure 5. Scores plot for creaminess in commercial yoghurts predicted by model based on experimental custard desserts, after calibration

Acknowledgement

We thank I. Polet, J.M. van Doorn, J. Prinz and L. Engelen for their assistance during this study.

References

1. Weenen, H.; van Gemert, L.J.; van Doorn, J.M.; Dijksterhuis, G.B.; de Wijk, R.A. *J. Text. Stud.* **2003**, *34*, 159-179.
2. Daget, N.; Joerg, M.; Bourne, M. *J. Text. Stud.* **1987**, *18*, 367-388.
3. Daget, N.; Joerg, M. *J. Text. Stud.* **1991**, *22*, 169-189.
4. Ward, C.D.W.; Stampanoni Koeferli, C.; Piccinali Schwegler, P.; Schaeppi, D.; Plemmons, L.E. *Food Qual. Pref.* **1999**, *10*, 387-400.

5. Elmore, J.R.; Heymann, H.; Johnson, J.; Hewett, J.E. *Food Qual. Pref.* **1999**, *10*, 465-475.
6. Richardson, N.J.; Stevens, R.; Walker, S.; Gamble, J.; Miller, M.; Wong, M.; McPherson, A. *Food Qual. Pref.* **2000**, *11*, 239-246.
7. Prentice, J.H. *J. Text. Stud.* **1973**, *4*, 154-157.
8. Wood, F.W. *Die Stärke* **1974**, *26* (4), 127-130.
9. Szczesniak, A.S. P. Sherman (Ed.). Academic Press, London, 1979; 1-20.
10. Kokini, J.L.; Cussler, E.L. *J. Food Sci.* **1987**, *48*, 1221-1225.
11. Cussler, E.L.; Kokini, J.L.; Weinheimer, R.L.; Moskowitz, H.R. 1979. Food texture in the mouth. *Food Technol.* 89-92.
12. De Wijk, R.A.; Van Gemert, L.J.; Terpstra, M.E.J.; Wilkinson, C.L. *Food Qual. Pref.* **2003**, *14*, 305-317.
13. Geladi, P.; Kowalski, B.R. *Anal. Chim. Acta* **1986**, *185*, 1-17.
14. Martens, H.; Naes, T. *Multivariate Calibration*; John Wiley & Sons, Chichester, 1989.
15. de Wijk, R.A.; Prinz, J.F.; Engelen, L.; Weenen, H. The role of alfa-amylase in oral texture perception. *Physiol. Behav.* **2004**, *83*, 81-91.
16. Weenen, H.; Jellema, R.H.; de Wijk, R.A. *Food Qual. Pref.* **2005**, *16*, 163-170.

Chapter 10

Effects of Structure Breakdown on Creaminess in Semisolid Foods

Hugo Weenen[1,2]

[1]Wageningen Centre for Food Sciences, Diedenweg 20, 6700 AN Wageningen, The Netherlands
[2]Current address: Numico R&D, P.O. Box 75338, 1118 ZN Schiphol Airport, The Netherlands

Saliva induced structure breakdown was found to be sufficiently rapid to affect sensory perception of starch containing semi-solid foods. Notably the attributes melting, thick, creamy, sticky, fatty, and some odour and flavour attributes were found to be altered when α-amylase was inhibited. As creamy is a particularly important attribute for the appreciation of a product, the effects of structure breakdown on creamy mouthfeel were studied in more detail. Interestingly, creaminess in a 3% fat containing dessert was not affected by the addition of an α-amylase inhibitor, whilst in a 0% fat containing dessert, creaminess increased by as much as 59%. To understand this, the structural evolution of an emulsion while in the mouth was studied by Confocal Scanning Laser Microscopy, and related to the sensory sub-attributes of creamy mouthfeel. This way, the different effects of salivary α-amylase in a 0% fat vs. a 3% fat dessert could be explained in terms of the effects of starch breakdown on roughness sensations, fat surfacing and the disintegrating structure of the continuous phase.

Creaminess has been the subject of several studies, in which the sensory and physical nature of creaminess has been investigated. Prentice (*1*) found that granularity or grittiness is a negative factor, which appears to be in agreement with a study by Wood (*2*), who suggested that creamy textured soups should have a very smooth mouthfeel, with complete absence of powderiness. Szczesniak (*3*) classified creamy in a category which she called feel on soft tissue surfaces, together with smooth and pulpy. Typical beverages having creamy characteristics included hot chocolate, eggnog and ice-cream soda, beverages typically not having creaminess characteristics included water, lemonade and cranberry juice.

Kokini *et al.* (*4*) found that creaminess can be predicted quantitatively based on three other sensory attributes, with the following formula: log creamy = 0.539log thick + 0.728log smooth + 0.220 log slippery. For a range of 16 experimental and commercial products chosen to have a wide range of rheological properties, R^2 was found to be 0.738. In fact they found that all texture attributes can be predicted from these three attributes, depending on the coefficients in the formula. Later studies with mainly dairy products indicated that creaminess can be predicted based on only two other attributes, i.e. thickness and smoothness (R^2=0.81; Kokini and Cussler, ref. *5*), according to the following formula: log creamy = 0.54log thick + 0.84log smooth. Clearly thickness and smoothness are important factors, however the accurateness of the creaminess prediction using this formula seems to depend on the choice of products, as a study with other products (*6*) resulted in a much lower R^2 (0.46). This led the authors to conclude that creamy is more than a combination of smoothness and thickness. Smoothness was related to the inverse of the sum of viscous and frictional forces, and thickness was related to shear stress (*7*).

Daget *et al.* (*8,9*) found a quadratic relationship between creaminess and the rheological parameters viscosity and flow behaviour index, in caramel creams and soups. Richardson and Booth (*10*) suggested that the contribution of dairy fat to creaminess perception is related to the contribution of the fat globules to the sensation of smoothness. It is suggested that the high density of even-sized small globules in homogenised milks or the high level of butterfat in cream, produce the sensation of smoothness, which together with the sensation of thickness generated by a sufficiently high viscosity, is necessary to give a realistic sensation of creaminess.

Clegg *et al.* (*11*) investigated the structural and compositional basis of creaminess in food emulsion gels (gelatin, mono-glyceride, skimmed milk powder, vegetable oil, sugar & water). Of four factors studied (oil droplet size, fat content, air bubble size and air content), especially oil droplet size and fat content were found to be important for creaminess.

De Wijk *et al.* (*12*) reported on creaminess in custard desserts varying in fat content, carrageenan and modified starch. Partial least square analysis of creamy mouthfeel based on sensory attribute scores showed positive regression values for creamy flavor, fatty flavor, thick mouthfeel, fatty mouthfeel and fatty afterfeel, whereas bitter chemical flavor, sickly flavor and rough mouthfeel had negative regression coefficients. Two latent variables explained 98% of the variance based on these sensory attributes.

Weenen et al. (13,14) report that in mayonnaises, custard desserts and starch based warm sauces, thick, airy, smooth and fatty (mouth- and/or afterfeel) contribute positively to creamy mouthfeel, while rough, heterogeneous, grainy and melting (mouth- and/or afterfeel) were found to be negative. Odor and trigeminal attributes had little or no effect on creamy mouthfeel, whereas flavour attributes did affect creamy mouthfeel, in some cases positively (caramel flavor) and in some cases negatively (broth & cheese flavor). Changing olfactory cues by using nose-clips or by adding a flavor, confirmed that aroma has an effect on creamy mouthfeel.

Roughness sensations have been related to friction (7) caused by a product when the tongue moves back and forth against the palate, and to vibrations in the skin resulting from the movement of a material over the skin (15). Roughness appears to be the opposite of smoothness, when evaluating the roughness and smoothness of textile (16).

Studies on the perception of semi-solid foods with added particles suggest that large, hard and sharp particles in a low viscosity medium produce a more rough, gritty and unpleasant sensation than small, soft and smooth particles in a higher viscosity medium (17-19).

For the study described here, the physiological and physical basis of the perception of rough mouthfeel is particularly relevant. The question whether roughness perception is caused by friction, the presence of particles or is due to vibrations, or a combination, may possibly be answered in the following way. Particles of the right size in a sample, will cause more friction, but also give rise to vibrations, when the tongue moves back and forth against the palate. The vibrations caused by the moving tongue could be the cue giving rise to roughness perception. The particles would also result in increased friction, caused by the sample, explaining the correlation between roughness perception and friction. To fully understand the relation between product properties and sensory perception, the fate of the product while being processed in the mouth has to be known, including the interaction of the evolving product + saliva mixture with the oral mucosa (20). A number of studies have addressed food – saliva interactions, and its relevance for sensory perception. The mixing of saliva with food can cause taste and flavor substances to become diluted (21, 22), or influence flavor perception in other ways (23-26). Saliva was also found to affect the compression and decompression forces relevant for sensory stickiness, suggesting that salivary α-amylase induced breakdown affects perceived stickiness of starch containing semi-solid foods (27). Recently a patent was filed in which the redistribution of fat globules is described, when starch containing foods are broken down by amylase during mastication (28). This phenomenon has been described as "fat surfacing", and is described in more detail in this paper.

De Wijk et al. (29) reported that boosting or inhibiting salivary α-amylase activity while eating/tasting modified starch containing custard desserts, strongly affected a number of sensory attributes, including creamy mouthfeel. The present study investigates the effects of structure breakdown on creamy mouthfeel, in particular how important sub-attributes of creamy mouthfeel (thick, melting, and rough moutfeel, flavor intensity) are affected by structure breakdown, and how this could affect creamy mouthfeel.

Materials and Methods

Effects of Inhibiting α-amylase on Sensory Ratings

Eighteen subjects, 11 females and 7 males aged between 18 and 34 yrs, participated the amylase study and 7 different subjects, all females aged between 39 and 56 yrs, participated in the acarbose study. The subjects participating in the latter study had been informed about acarbose and its possibly side effects if swallowed. To avoid these side effects, subjects expectorated rather than swallowing the acarbose samples. All subjects signed consent forms expressing their willingness to participate in the studies, and all had previously been screened for olfactory and taste disorders and had received extensive training in the description of sensory mouthfeel and afterfeel attributes for semi-solid foods. The subjects were paid for their participation. Testing took place at the sensory facilities of TNO-Nutrition and Food Research in Zeist, The Netherlands. Two commercially available starch contaning vanilla custard desserts were used to study the effects of amylase induced structure breakdown and fat on sensory perception (0.1% fat Creamex vla, Creamex, Rijkervoort, The Netherlands; and 3% fat Boerenland vla, Campina, Woerden, The Netherlands). Full details of the study have been described elsewhere (29).

Sensory Time Intensity Measurements

The sensory time intensity panel consisted of six selected and experienced assessors. Training was done during two sessions of two hours each with both model and commercial samples. For the measurements the following procedure with the computer system FIZZ was applied. All six samples were judged on all five attributes twice during two sessions of approximately two hours (thus in total 2 x 2 = 4 replications), i.e. per session assessments of six samples, a break (15 minutes), and again assessments of six samples (thus, per session in total 6 samples x 5 attributes x 2 replications = 60 spoons had to be taken in the mouth).

Serving order was randomized per assessor and per replication. Samples were stored in the refrigerator and served immediately after removing a sample from the refrigerator. Per sample approximately 75 g was served in a plastic cup. For each attribute an unused plastic tablespoon was available of which a flat spoonful (approximately 10 g) had to be taken in the mouth for evaluation. Sequence of attributes (Dutch term between brackets): 1. Cold (Koud), 2. Creamy (Romig), 3. Fatty (Vettig), 4. Sticky (Plakkerig) and 5. Astringent (Stroef). Anchors for respectively bottom and top of vertical bar were for cold: warmer and colder, and for all the others: very little and very much. Assessments started as soon as all material from the spoon was put into the mouth. Subjects were instructed and trained to swallow the product 20 seconds after the custard dessert was put into the mouth. Total evaluation time per attribute was 60 seconds. A zero rating was defined as the intensity of the attribute based for one's own saliva. Water was available for rinsing the mouth, in between evaluations.

Samples used for the time intensity study were prepared by UHT treatment at NIZO, Ede, the Netherlands, were preheated to 75°C, indirectly heated for 5 seconds at 144°C, and then cooled to below 5°C. All samples contained 4.5% starch (VA85T), 6.5% sucrose, no carrageenan, and 0.1% vanilla powder (Danisco, code 3912). Fat contents varied from 0-15%.

Microstructural analysis by confocal scanning laser microscopy

CSLM instrumentation and observation conditions. Observations were done with a LEICA TCS SP Confocal Scanning Laser Microscope (CSLM), in single photon mode, configured with an inverted microscope (model Leica DM IRBE), and using an Ar/Kr laser. The following Leica objective lenses were used: 20x/0.7NA/dry/HC PL APO, 63x/UV/1.25NA/water immersion/PL APO. Fluorescent probes used were FITC (fluorescein-5-isothiocyanate) and Nile Red. The excitation wavelengths were 488 and 568 nm, respectively, and the emission wavelength were at 525 and 525-605 nm. Observation was done at the outer surface of the product at locations away from the surface of the cover glass, prevent any effects from possible glass/product interactions. Digital image files were acquired in multiple .tif format and in 1024x1024 pixel resolution.

Non-covalent labelling. A mixture of FITC with Nile Red was used for non-covalent labelling. A small piece of sintered glass containing in its pores an amount of two drops of a 0.1% FITC and 0.01% Nile Red in polyethylene glycol-glycerol solution was added to 2 ml of sample. The probe molecules will spread from the pores of the sintered glass over the sample according to local accessibility and affinity. The applied probes show different affinities for hydrophilic and hydrophobic domains. The FITC molecules accumulate in hydrophilic regions and the Nile Red molecules accumulate in hydrophobic regions. Consequently, structural elements will be highlighted in a way that is dominated by the kinetics and thermodynamics of the probe molecules in the specimen matrix.

Mouth processing. A spoonful of product was taken directly from the refrigerator and put into the mouth. When the product was in the mouth, the tongue was moved back, forth and sideways against the palate in a controlled way, moving the dairy custard around as well. Subjects were trained prior to experiments. After approximately 10 to 15 sec. of oral processing the product was spat out into a small container. A sample volume of 1.5 ml was transferred into a small Eppendorf tube, on top of 10 µL FITC/NileRed solution, followed by very mild and gentle mixing. From the stained product a very small sample was taken and transferred onto a cover glass followed by observation in the microscope. The total time between taking the sample from the refrigerator and actual observation in the CSLM was less then 1 minute. Validation experiments showed that longer preparation times resulted in identical findings.

Video CSLM. A droplet of a commercial 3% fat custard dessert (Albert Heijn, vanille vla) was carefully mixed with about 2% FITC 0.1%, and Nile Red (0.01%). A droplet was added to the CSLM, carefully smeared over the glass slide and 1-2 photographs were taken. A droplet of fresh human saliva was added, and a series of photographs were taken with a resolution of 512-1024 dpi.

Results and Discussion

Fat surfacing

In order to investigate whether and to what extent oral processing affects the structure of semi-solid foods and ultimately their sensory perception, we examined the structural evolution of starch containing custard desserts during oral processing. For this purpose trained subjects were asked to keep a 5 ml sample of a commercial 3% fat containing custard dessert in the mouth, to keep it there while manipulating the sample with the tongue, and to spit it out after 10-15 seconds. The spat out sample was immediately processed and examined by confocal scanning light microscopy (CSLM). Fluorescent dyes were added to distinguish between the aqueous and lipophilic phases. Two clear structural differences occurred as a result of oral processing: 1. the structures of the starch granules disappeared; 2. the fat globule distribution had changed after oral processing. In the original intact sample fat globules, visualised as red dots, were homogeneously distributed over the entire sample (Figure 1a), however in the sample which had been in the mouth, the distribution of the fat globules had become inhomogeneous, with a higher concentration of fat globules at the surface (Figure 1b) and around air bubbles (Figure 1c). These observations raised a number of questions: 1. Is saliva induced starch breakdown of semi-solid foods fast enough to be relevant for events taking place while eating, and ultimately for sensory perception? 2. If yes, then how would this affect sensory perception? 3. What would be the mechanism for this redistribution of fat globules, or "fat surfacing"?

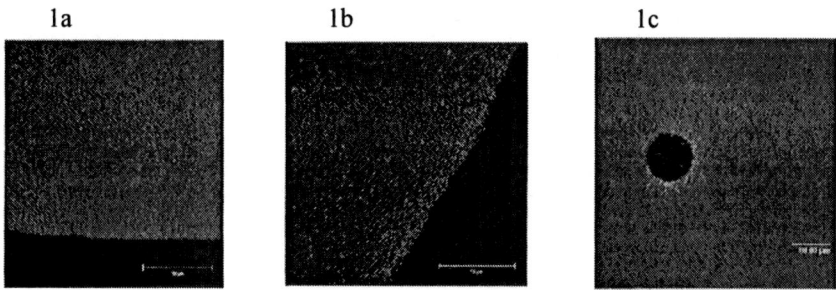

Figure 1. CSLM pictures of starch containing dessert before oral processing (1a) and after oral processing (1b & 1c) (See page 1 in color insert.)

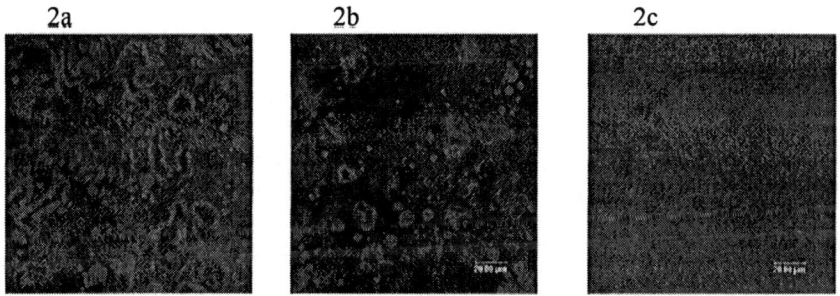

Figure 2. CSLM pictures of a 3% fat custard dessert before adding saliva (2a), and 15 seconds (2b) and 120 seconds (2c) after adding saliva (See page 1 in color insert.)

Mechanism of fat surfacing

In an intact starch based dairy emulsion, the fat globules are kept in place by the protein and starch (amylose and amylopectin) network, which prevents the fat globules from moving upward (creaming) and from coalescing. When the starch network is broken down by α-amylase, viscosity drops and the fat globules are more free to move and can move upward as in creaming, resulting from the effects of gravitation (Figure 1b, 2c & 3). However in a sample that is manipulated in the mouth, two other phenomena take place, which may accelerate fat surfacing. One such phenomenon is aeration, or the mixing of air

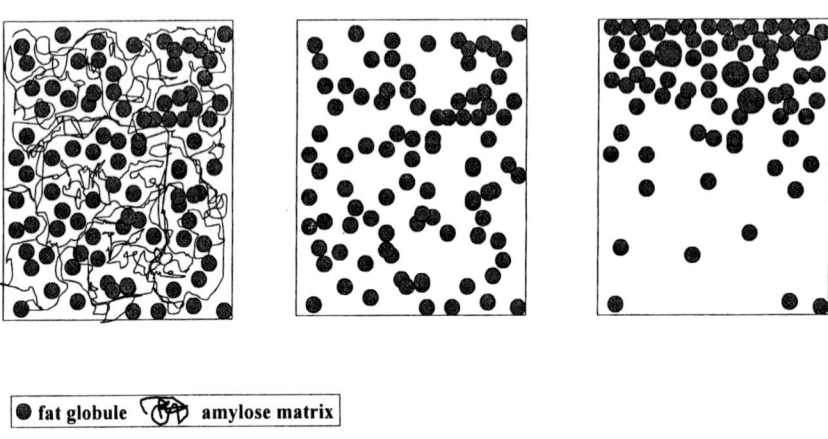

Figure 3. Schematic representation of saliva induced breakdown of starch network and fat surfacing

bubbles with the sample. Air is apolar, and so is fat, therefore fat globules tend to adhere to air bubbles. Figure 1c is a CSLM picture of an air droplet surrounded with fat globules taken of a spat out custard dessert sample, and clearly shows that air droplets may indeed accelerate fat surfacing. This phenomenon is schematically depicted in Figure 4.

Another phenomenon occurring during oral processing but not under static conditions is surface renewal. When manipulating a product in the mouth, the surface of the sample which is in contact with air will be deformed, hence the surface will expand and contract, in other words the surface is dynamic and is constantly renewed. As fat globules preferentially adhere to air – aqueous phase interfaces, the surface renewal process will facilitate fat surfacing, analogous to air bubbles mixing with the product.

Inhibition of α-amylase activity using an α-amylase inhibitor

To determine whether starch breakdown in the mouth is fast enough to affect sensory perception of semi-solid foods, a sensory panel evaluated custard desserts with various concentrations of added acarbose, an α-amylase inhibitor

(*29*). The results indicated that creamy mouthfeel ratings were significantly affected in a 0% fat product (Figure 5, p=0.001), but not in a 3% fat product (p=0.73). Of the other mouthfeel attributes, cold, sticky, melting and fatty moutfeel were significantly affected in both the 0% and 3% fat desserts, thick and dry/mealy only in the 3% fat dessert. As creaminess ratings generally go up with increasing fat contents (*12*), we had expected that fat surfacing would increase creamy ratings, hence that inhibiting fat surfacing by inhibiting starch breakdown, would decrease creamy ratings. Therefore, based on only fat surfacing, creaminess ratings of the 3% fat dessert were expected to decrease when α-amylase is inhibited, whilst creaminess ratings of the 0% fat product should not be affected. However creamy mouthfeel is not only affected by the presence of fat (at the surface). Other factors known to affect creamy mouthfeel include the rheological properties of the bulk phase of an emulsion (*4,5,8,9,13,14,31*), and subjective roughness (*13, 14, 31, 32*).

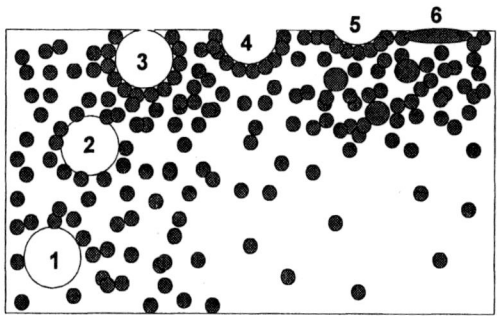

Figure 4. Schematic representation of fat surfacing by air bubbles

Roughness (and grittiness) intensities are affected by particles size, hardness and sharpness, but also by the viscosity of the medium (*17-19*). This results in a complex of cooperating and opposing factors affecting creamy mouthfeel. Scheme 1 shows how the factors described above interact and affect creaminess in a starch containing semi-solid food. In summary, the main factors relevant for creaminess ratings of starch containing semi-solid foods under mastication conditions include:

1. Subjective roughness caused by starch particles, which break down when exposed to α-amylase containing saliva. Particles of sizes >2 μm have been shown to increase rough mouthfeel and decrease creamy mouthfeel (*19*). In fact

the largest effects of added particles on texture attributes were observed for particles around 70-80 μm in diameter, which is about the size of swollen starch particles. As starch breaks down when in contact with saliva, subjective roughness is expected to decrease, and creaminess to increase.

2. Lubrication and flavour release by fat globules, which are expected to increase as a result of fat surfacing (see above, and ref. *28*), when the starch network disintegrates. Fat surfacing can of course not occur in a 0% fat product.

3. Subjective thickness and melting caused by the starch network, which breaks down when a food is exposed to α-amylase containing saliva. Melting is defined by sensory panellists as "The rate at which a product becomes liquid and spreads in the mouth...." (*33*), therefore is expected to increase when starch is broken down. As expected, when starch breakdown is inhibited by an α-amylase inhibitor (*29*), the melting mouthfeel attribute was found to decrease most strongly, and thick mouthfeel was found to increase. As creamy mouthfeel is positively correlated with thickness ratings, and negatively with melting ratings (*12-14*) these effects are expected to increase creamy mouthfeel ratings.

4. Subjective grittiness resulting from particles present in a semi-solid dispersion is higher in lower viscosity media than in higher viscosity media (*17, 18*). Similarly, it can be expected that subjective roughness resulting from particles in a dispersion increases with decreasing viscosity (*19*). When salivary

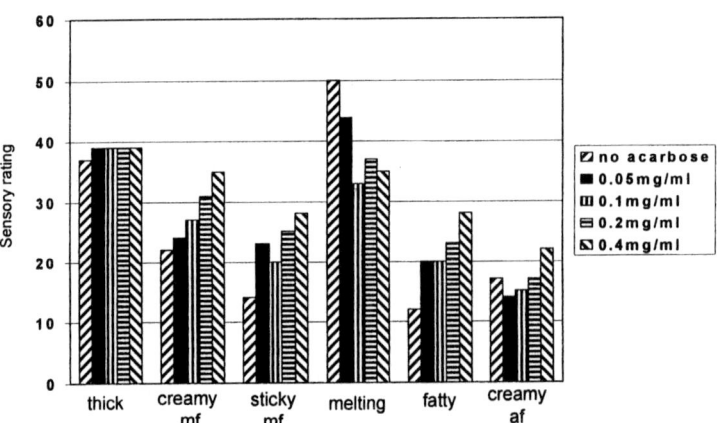

Figure 5. Effects of α-amylase inhibition on texture related sensory attributes in 0% fat starch containing dessert (mf: mouthfeel; af: afterfeel). Creamy mouthfeel, melting mouthfeel, sticky mouthfeel and fatty afterfeel were significantly affected by α-amylase inhibition, thick mouthfeel, creamy afterfeel and other texture attributes (not shown) were not significantly affected.

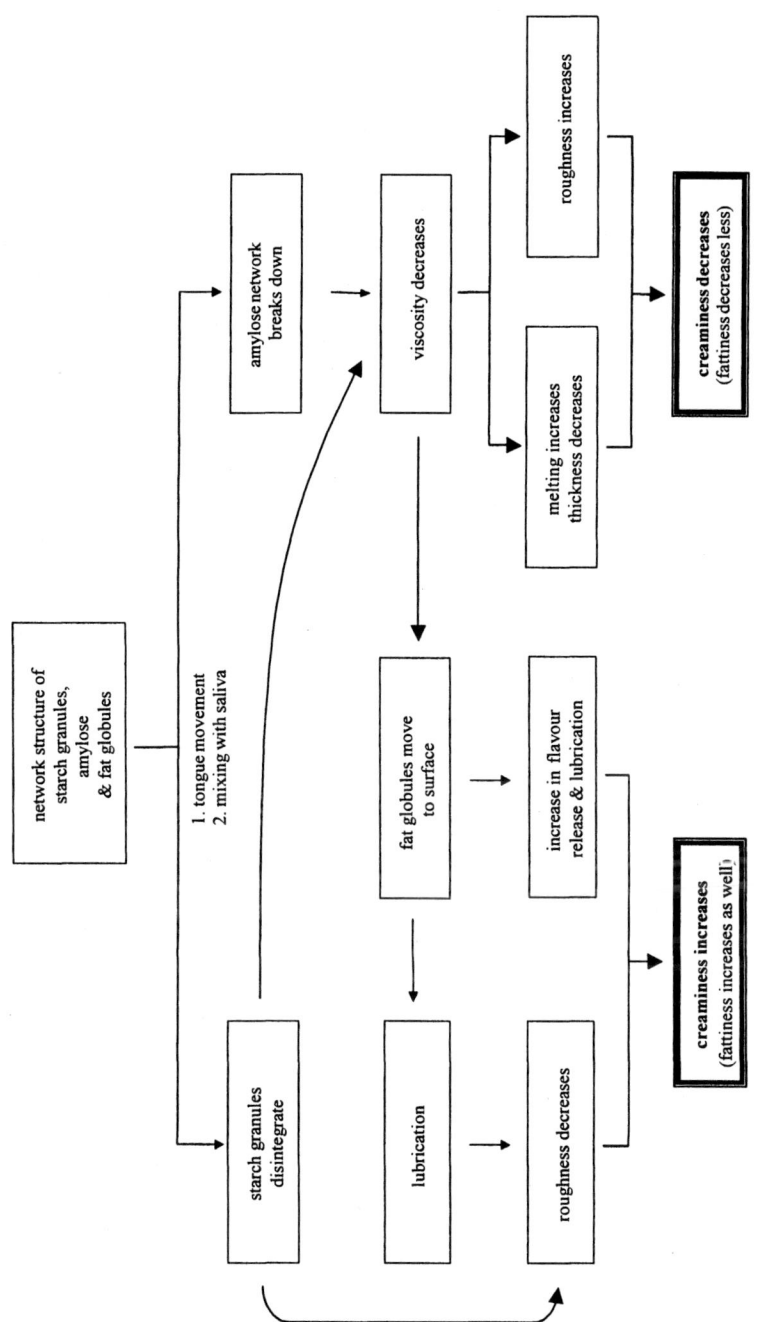

Scheme 1. *Factors affecting creamy mouthfeel in (starch containing) semi-solid foods*

amylase breaks down starch, the viscosity of the medium decreases, and therefore roughness is expected to increase.

In addition the rough surface of the tongue may contribute to sensations of roughness. Especially the dorsal surface of the posterior two thirds of the human tongue is relatively rough, and can be expected to play a role in subjective roughness. In fact time intensity measurements of sensory roughness (Figure 6) seem to confirm this. After swallowing, when the product remains on the tongue are disappearing, sensory roughness was found to increase temporarily, before continuing to decrease, when the product is broken down further by salivary amylase, and eventually disappears. This can best be explained by the subjects sensing of the rough surface of its own tongue, and attributing this sensation to the product that is under evaluation. Roughness would then be enhanced if the viscosity of the product decreases as a result of starch breakdown by α-amylase.

Upon starch breakdown in the mouth, factors 1&2 are expected to cause creaminess to increase, while factors 3&4 are expected to decrease creaminess (Scheme 1). Whether the net effect is positive or negative depends among others on the fat content of the sample, because, when no fat is present, fat surfacing can obviously not take place. This explains why creaminess ratings of the 0% fat custard dessert in this study increased when α-amylase is inhibited, while creaminess ratings of the 3% fat sample do not show this effect.

Surprisingly, inhibition of amylase increased thickness only in the 3% fat dessert, not in the 0% fat dessert, suggesting that perceived thickness is affected by factors other than just viscosity averaged over time.

The structure of starch containing foods is broken down in the mouth, mainly because of amylase induced breakdown. However the factors affecting creaminess described here, may also be applicable to other structure breakdown mechanisms, such as shear and elongational stress induced structure breakdown, and structure breakdown resulting from dilution of a product with saliva.

Acknowledgements

We would like to gratefully acknowledge J. Klok (NIZO) and M. Paques (FCDF) for the microstructural work which led to the fat surfacing hypothesis. In addition we are grateful for the contributions of J. Bergsma (Avebe), A.M. Janssen (WCFS), R.A. de Wijk (WCFS) and R. Hamer (WCFS) for their very useful suggestions and comments during the course of our investigations.

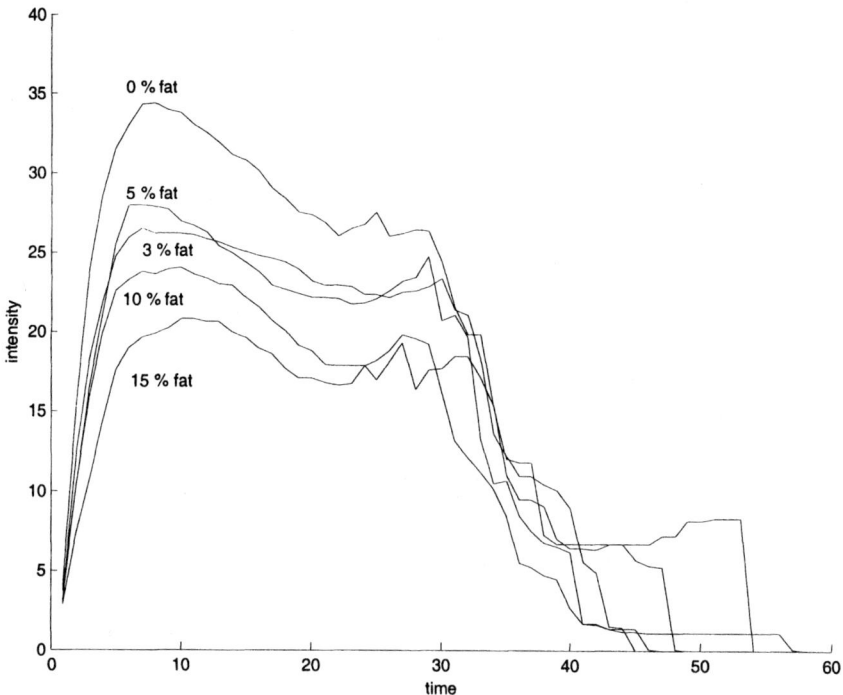

Figure 6. Mean rough mouthfeel ratings from a sensory panel (N=6) for desserts varying in fat percentage (0-15%). Products were swallowed after 20 seconds

References

1. Prentice, J.H. *J. Text. Stud.* **1973**, *4*, 154-157.
2. Wood, F.W. *Stärke* **1974**, *26* (4), 127-130.
3. Szczesniak, A.S. Classification of mouthfeel characteristics of beverages. In: *Food texture and rheology,* Sherman, P., Ed.; Academic Press, London, **1979**; 1-20.
4. Kokini, J.L.; Kadane, J.; Cussler, E.L. *J. Text. Stud.*, **1977**, *8*, 195-218.
5. Kokini, J.L.; Cussler, E.L. *J. Food Sci.* **1987**, *48*, 1221-1225.
6. Cussler, E.L.; Kokini, J.L.; Weinheimer, R.L.; Moskowitz, H.R. *Food Technol.* **1979**, 89-92.
7. Kokini, J.L. *J. Food Eng.* **1987**, *6*, 51-81.
8. Daget, N.; Joerg, M.; Bourne, M. *J. Text. Stud.* **1987**, *18*, 367-388.
9. Daget, N.; Joerg, M. *J. Text. Stud.* **1991**, 22: 169-189.

10. Richardson, N.J.; Booth, D.A. *J. Sens. Stud.* **1993**, *8*, 133-143.
11. Clegg, S.; Kilcast, D.; Arazi, S. In *Proceedings of third International Symposium on Food Rheology and Structure* **1993**, pp. 373-377.
12. De Wijk, R.A.; Van Gemert, L.J.; Terpstra, M.E.J.; Wilkinson, C.L. *Food Qual. Prefer.* **2003**, *14*, 305-317.
13. Weenen, H.; Jellema, R.H.; de Wijk, R.A. *Food Qual. Prefer.* **2005**, 16(2), 163-170.
14. Weenen, H.; Jellema, R.H.; de Wijk, R.A. In: *Flavour and texture of lipid containing foods*; Weenen, H., Shahidi, F. Eds., American Chemical Society, Washington DC, **2005**, this volume.
15. Christensen, C.M. In: *Advances in food research*, Vol. 29. Chichester C.O. editor. Academic Press Inc., London, UK **1984**, pp. 159-198.
16. Stevens, S.S.; Harris, J.R. *J. Exp. Psychol.* **1962**, *64*, 489-494.
17. Imai, E.; Hatae, K.; Shimada, A. *J. Text. Stud.* **1995**, *26*, 561-576.
18. Imai, E.; Shimichi, Y.; Maruyama, I.; Inoue, A.; Ogawa, S.; Hatae, K.; Shimada, A. *J. Text. Stud.* **1997**, *28*, 257-272.
19. Engelen, L.; De Wijk, R.A.; Van der Bilt, A.; Prinz, J.F.; Janssen, A.M.; Bosman, F. Relating particles and texture perception. *Physiol. Behav.* **2005**. (In press).
20. Wilkinson, C.; Dijksterhuis, G.B.; Minekus, M. *Trends Food Sci. Technol.* **2001**, *11*, 442-450.
21. Ruth, S.M.V.; Roozen, J. P.; Nahon, D. F.; Cozijnsen, J. L.; Posthumus, M. A. *Z-Lebensm-Unters-Forsch.* **1996**, *203*, 1-6.
22. Christensen, C.M. In: *Clinical measurements of taste and smell*. Meiselman H.L., Rivlin, R.S. Eds. MacMillan, New York, NY, **1985**, pp. 414-428.
23. Haring, P.G.M. In: *Flavour science and technology*. Bessiere, Y.; Thomas, A. F. (editors). Wiley, Chichester, **1990**, 351-354.
24. Ruth, S.M.v.; Roozen, J.P. *Food Chem.* **2000**, *71*, 339-345.
25. Harrison, M. *COST Action.* **1998**, 2, 91-96.
26. Guinard, J.-X.; Zoumas-Morse, C.; Walchak, C.; Simpson, H. *Physiol. Behav.* **1997**, *61*, 591-596.
27. Dunnewind, D.; Janssen, A.M.; van Vliet, T.; Weenen, H. Accepted for publication in *J. Text. Stud.* **2005** (In press).
28. Paques, M.; Weenen, H.; van Riel, J.; Engelen, L.; Hamer, R. Evaluation of in mouth sensory properties. EPO01202438.6, **2001**.
29. De Wijk, R.A.; Prinz, J.F.; Engelen, L.; Weenen, H. The role of alpha-amylase in oral texture perception. Physiol. Behav., **2004**, *83*, 81-91.
30. Janssen, A.M.; Terpstra, M.E.J.; De Wijk, R.A.; Prinz, J.F. Unpublished results.
31. De Wijk, R.A.; Prinz, J.F. *Food Quality Pref.* **2005**, *16,* 121-129.
32. De Wijk, R.A.; Prinz, J.F.; Weenen, H. *J. Food Sci.* **2005**, (Submitted).
33. Weenen, H.; van Gemert, L.J.; van Doorn, J.M.; Dijksterhuis, G.B.; de Wijk, R.A. *J. Text. Stud.* **2003**, *34*, 159-179.

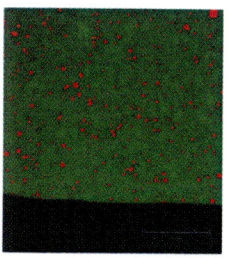 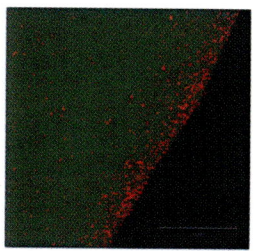 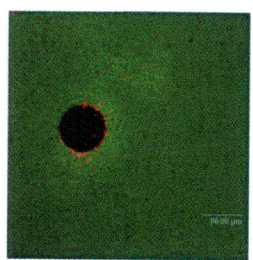

Figure 1. *CSLM pictures of starch containing dessert before oral processing (1a) and after oral processing (1b & 1c)*

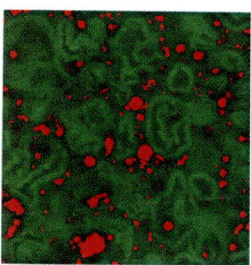 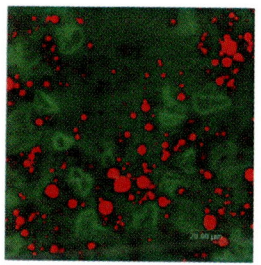 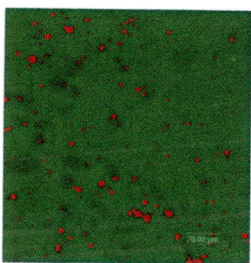

Figure 2. *CSLM pictures of a 3% fat custard dessert before adding saliva (2a), and 15 seconds (2b) and 120 seconds (2c) after adding saliva*

Chapter 11

Chemistry and Rheology of Cheese

Michael H. Tunick and Diane L. Van Hekken

Dairy Processing and Products Research Unit, Eastern Regional Research Center, Agricultural Research Service, U.S. Department of Agriculture, 600 East Mermaid Lane, Wyndmoor, PA 19038

The chemistry of cheese and the use of this knowledge to develop a low-fat Mozzarella, now being used in the National School Lunch Program, were investigated. Texture and flavor development in cheese result from complex interactions between fat and protein, which are influenced by other factors such as salt content, pH, and temperature. Enzymes from the rennet and the starter cultures cleave casein molecules resulting in microstructural changes in the curd and forming compounds that impart specific flavors to the cheese. Fat serves as a filler in the protein matrix and as a carrier of flavor; consumers often find low-fat cheese varieties to be harder, less meltable, and less flavorful. These quality and sensory problems are surmountable if manufacturing procedures are adjusted with consideration of the chemistry taking place. The flavor can be enhanced by adding certain adjunct cultures, and the texture and meltability can be improved by processing at lower temperatures, which allows the microorganisms to remain active and break down the protein matrix during storage.

The texture and flavor of cheese arise from chemical interactions which are controlled by manufacturing and ripening procedures. The development of a product such as low-fat Mozzarella is a result of a thorough knowledge of the chemistry of cheese as evaluated by rheology, microscopy, and other techniques. Ripened cheese is made from milk that has been fermented by bacteria and coagulated by an enzyme, producing a curd which is separated from whey by draining and pressing. After salting, the cheese is stored under specific conditions to cause changes in rheology and flavor as brought about by the action of microflora, enzymes, and chemical compounds. The manufacturing parameters and ripening conditions are responsible for the roughly 2000 different varieties of cheese in the world (1).

Cheese Chemistry

Lipids

As with most foods, lipids in cheese consist almost exclusively of triacylglycerols. About 2% of the lipid consists of monoacylglyceols, diacylglycerols, and free fatty acids (2). Lipids are responsible for a significant portion of the rheological properties of cheese because they serve as filler in the protein matrix. Viscosity is a major factor in detecting fat in the mouth; if viscosity is increased by homogenization or fatty acid saturation, the awareness of fat is heightened (3). The perception of fat content is also enhanced as it melts and cools at body temperature (3).

Lipids are also largely responsible for cheese flavor, as they carry hydrophobic flavor compounds produced from proteolysis. In addition, milkfat contains naturally occurring flavors, which develop through interactions with heat, light, oxygen, and food components. Lipolysis also leads to flavor compound formation, especially in blue mold cheeses where *Penicillium roqueforti* and *P. camemberti* secrete lipases (4). Lipolytic breakdown of triacylglycerols leads to formation of keto acids, hydroxyacids, and fatty acids, which are further metabolized into methyl ketones, thioesters, esters, alcohols, and lactones (4). Lipids modify the perception of flavors, including their timing and release (5).

Protein

Protein provides structure to cheese, and proteolysis during ripening leads to softening, increased meltability, and formation of flavor compounds. Almost all protein in cheese consist of casein. There are four main types of casein found in bovine milk: α_{s1}-casein (about 38% of the total casein), α_{s2}-casein (10%), ө-casein (34%), and ϭ-casein (15%), all with molecular weights between 20 and 25 kDa (2). Caseins are arranged in submicelles, around 5000 kDa molecular weight and with ϭ-casein primarily on the surface. Submicelles combine to form micelles, with a molecular weight of 1.3×10^6 kDa and a diameter of 50-500 nm (2). Cleavage of ϭ-casein by the chymosin coagulant causes the micelles to destabilize and aggregate into a gel under pressure, with heat, as whey is removed. The gel eventually fuses, becoming the cheese curd.

Ripening

The characteristic flavor and texture of cheese arise from the metabolic processes of specific microorganisms. Some flavor is generated through fermentation pathways of the bacteria or by Maillard, Strecker, and other chemical reactions. However, most cheese flavor is brought about by the action of enzymes indigenous to the milk (mainly plasmin), added to the milk (chymosin), or released by lysed bacterial cells. Bacterial ripening of cheese is affected most by water activity, NaCl concentration, pH, and temperature. Water hydrated to casein is bound, and bacteria survive in the unbound water. The NaCl concentration is between 0.7 and 7% in cheese, which is enough to inhibit, but not stop, bacterial growth. The pH of cheese is usually between 4.5 and 5.3; most bacteria do not grow below pH 5.0. The ripening temperature can be between 5°C and ambient, and is the cheesemaker's primary method of controlling ripening (2).

The rheology of cheese changes as proteolytic breakdown of the casein, particularly α_{s1}-casein, progresses. Figure 1 shows how the microstructure of full-fat Mozzarella is altered during 6 wk of storage at 4°C. At the outset, the casein matrix is interrupted by fat globules, some of which have starter culture bacteria at the fat-protein interface. The matrix is degraded with time, allowing the globules to coalesce, and the bacteria to lyse and disappear (6).

Transmission electron micrographs of casein submicelles in the same Mozzarella sample show that rearrangement takes place during storage (Figure 2). The proteolysis of α_{s1}-casein, the primary structural protein in cheese, and the concurrent formation of peptides apparently facilitates the reorganization of submicelles. The original homogeneous pattern, where the average spacing is 17 nm, is transformed into a pattern of clusters, where the average spacing is 40 nm. This result may provide the molecular basis for the observed weakening of the casein matrix (5).

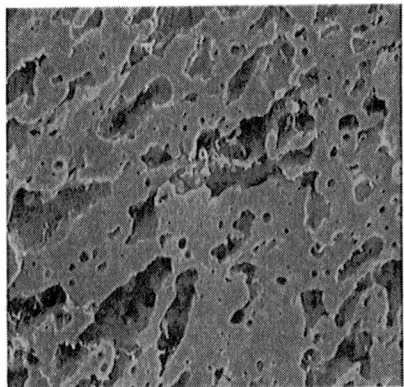 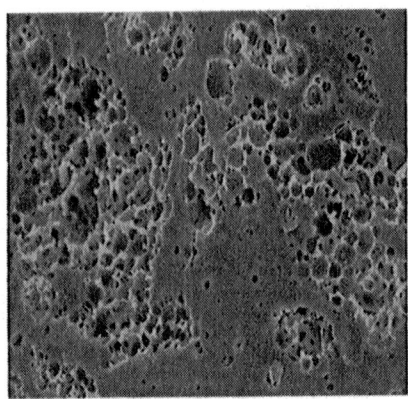

Figure 1. Scanning electron micrographs of full-fat Mozzarella cheese at 0 wk (left) and 6 wk (right) of storage at 4 °C, showing changes in fat globules (dark cavities) within the casein matrix (smooth areas). Scale bar at lower right corresponds to 40 µm.

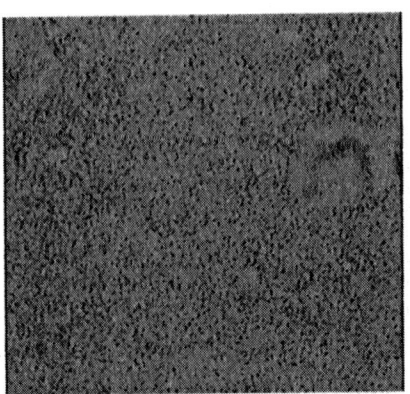 

Figure 2. Transmission electron micrographs of full-fat Mozzarella cheese at 0 wk (left) and 6 wk (right) of storage at 4 °C, showing rearrangement of casein submicelles (dark spots) from a homogeneous pattern to clusters. Scale bar at lower right corresponds to 100 nm.

Cheese Rheology

Rheology is the study of the flow and deformation of matter, whereas texture is a composite sensory attribute referring to flow, deformation, and disintegration under force (8). A trained panel is often used to quantify the texture of cheese, but instruments are required for rheological investigations. Chemical processes result in bond breaking, movement of fat globules and casein submicelles, and disruption of structure, all of which affect the rheological properties of cheese (9).

The three categories of rheological measurements are empirical, imitative, and fundamental. Empirical measurements have been reviewed by Szczesniak (10) and include such tests as penetration, ball compression, and curd tension. These tests measure parameters that are related to texture, but the experimental conditions are arbitrary and the results are not comparable to those from more rigorous tests (11).

Texture Profile Analysis (TPA)

Imitative measurements are obtained by using universal testing machines to perform texture profile analysis (TPA), which was first applied to foods by Bourne (12). TPA is performed by sending a crosshead vertically downward onto a specimen, compressing it twice between parallel plates to mimic the action of biting twice on a piece of cheese. The maximum force obtained on the first compression is defined as hardness, the ratio of the positive force area of the second compression to that of the first is cohesiveness, the height the specimen recovers between compressions is springiness, and these three quantities multiplied together is chewiness (8).

Uniaxial Compression

Fundamental measurements include compression, dynamic, transient, and torsion tests. In the simplest test, uniaxial compression, a universal testing machine is used to apply a stress downward on a specimen. The resulting deformation is measured as the ratio of height change to original height (engineering strain). A stress-strain curve can be used to determine stress and strain at fracture and work performed up to fracture (13).

Dynamic Tests

Dynamic measurements are made using small amplitude oscillatory shear, where a specimen disk is placed between two parallel plates and subjected to sinusoidal oscillation, with either stress or strain being varied harmonically with time. The experiment must be performed in the linear viscoelastic region to keep the structure intact. The quantities obtained are GN, the elastic or storage modulus, which is a measure of solid-like behavior; GO, the viscous or loss modulus, a measure of fluid-like behavior; and $\square^*$, the complex viscosity, a measure of viscoelastic flow (8).

Torsion Tests

In a torsion test, a capstan- or hourglass-shaped specimen is twisted until it breaks in the narrow center, and the shear stress and shear strain at fracture are noted. This test was first used on food at North Carolina State University (14) and has recently been applied to cheese (15). A texture map, a graph of torsion shear stress vs. torsion shear strain, provides a picture of how the characteristics of cheese change with age (16). Figure 3 shows that Romano, a hard cheese, turns brittle as it ripens, Mozzarella, which is stretchable, becomes more rubbery, and Cheddar, a semi-hard cheese, initially becomes tougher but then becomes mushier as proteolysis continues. Brick, Colby, and Havarti also become more rubbery. Gouda becomes mushier, but Old Amsterdam, which is aged without protective packaging, turns brittle. These changes are related to composition and proteolysis (17).

Transient Tests

Transient tests can be performed in compression, shear, torsion, and other modes. In a stress relaxation test, a sudden and constant strain is applied on a specimen and the resulting stress is measured over time. The ratio of stress to strain is the relaxation modulus. Conversely, in a creep recovery test, a sudden and constant stress is applied and the resulting strain is measured over time. Creep compliance, the ratio of strain to stress, is obtained. Transient tests are conducted at higher strains than dynamic tests, causing breakage of bonds and furnishing information on long-range properties of the casein network (8).

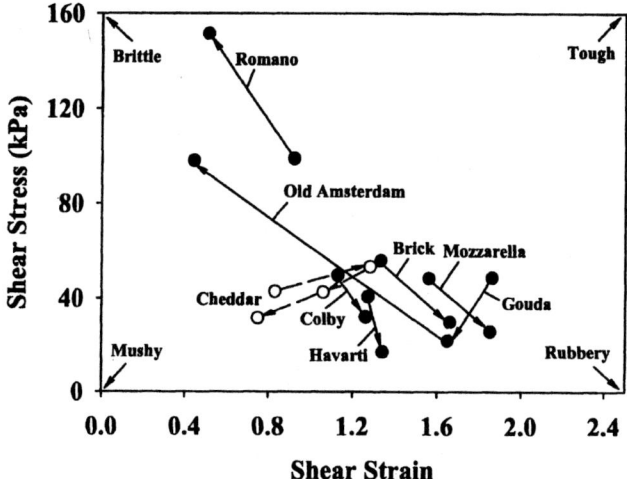

Figure 3. Texture map of various cheeses, showing changes in characteristics during aging. Arrows show direction of change for each variety from fresh to ripened.

Application: Low-Fat Mozzarella

Insights gained by cheese chemistry and rheology are applicable to the development of new or improved products. For instance, in 1992 the Food and Nutrition Service of USDA requested a reduced-fat Mozzarella for pizza lunches served in the National School Lunch Program. Pizza is the favorite lunch in schools and the Mozzarella topping is an important protein and calcium source, but the cheese normally contains 17-24% fat. Commercial low-fat cheeses available at the time displayed hard texture, little flavor, and minimal melting (18).

Research by the Dairy Processing and Products Research Unit demonstrated that peptides produced by proteolysis of $_{s1}$-casein help break up the protein matrix, leading to a softer and more meltable cheese (19). Experimental manufacturing runs were conducted using lower cooking and stretching temperatures than usual, enabling enzymes and starter culture bacteria to remain active longer. The storage time was increased to 4-6 wk, allowing more proteolysis to take place. An adjunct culture was also used with the starter to increase flavor. Electron microscopic imaging revealed that the microstructure of the cheese changed dramatically during storage, with degradation of the casein network (as confirmed by electrophoresis) and coalescence of the fat globules (leading to increased meltability) (20). TPA and dynamic measurements showed that the hardness, springiness, cohesiveness, GN, and GO values of the cheese were similar to those of part skim Mozzarella.

The resulting cheese, which contained 9-12% fat, was accepted by students in public school trials (21) and adopted by the National School Lunch Program in 1995. Reduced-fat Mozzarella has been served in American schools since then, the result of careful and systematic studies of cheese chemistry and rheology.

Acknowledgments

The authors thank Eleanor M. Brown, Peter H. Cooke, Harold M. Farrell, Jr., Virginia H. Holsinger, Thomas F. Kumosinski, Edyth L. Malin, James J. Shieh, and Philip W. Smith for their contributions.

Mention of trade names or commercial products is solely for the purpose of providing specific information and does not imply recommendation or endorsement by the U.S. Department of Agriculture.

References

1. Kosikowski, F.V. *Sci. Am.* May 1985, p 88.
2. Fox, P.F.; Guinee, T.P.; Cogan, T.M.; McSweeney, P.L.H., Eds.; *Fundamentals of Cheese Science;* Aspen Publ.: Gaithersburg, MD, 2000.
3. Mela, D.J.; Langley, K.; Martin, A. *Appetite* **1994**, *22,* 67-81.
4. Fox, P.F. In *Encyclopedia of Dairy Sciences;* Roginski, H., Fuquay, J.W., Fox, P.F., Eds.; Academic Press: Amsterdam, 2003; Vol. 1, pp 320-326.
5. Miettinen, S.M.; Hynönen, L.; Tuorila, H. *J. Agric. Food Chem.* **2003**, *51,* 5437-5443.
6. Tunick, M.H.; Shieh, J.J. In *Chemistry of Structure-Function Relationships in Cheese*; Advances in Experimental Medicine and Biology, #367; Malin, E.L., Tunick, M.H., Eds.; Plenum Press: New York, NY, 1995; pp 7-19.
7. Tunick, M.H.; Cooke, P.H.; Malin, E.L.; Smith, P.W.; Holsinger, V.H. *Int. Dairy J.* **1997**, *7,* 149-155.
8. Tunick, M.H. *J. Dairy Sci.* **2000**, *83,* 1892-1898.
9. Gunasekaran, S.; Ak, M.M., Eds.; *Cheese Rheology and Texture;* CRC Press: Boca Raton, FL, 2003.
10. Szczesniak, A.S. *J. Food Sci.* **1963**, *28,* 410-420.

11. Rao, V.N.M.; Skinner, G.E. In *Engineering Properties of Foods;* Rao, M.A., Rizvi, S.S.H., Eds.; Marcel Dekker: New York, NY, 1986; pp 215-254.
12. Bourne, M.C. *J. Food Sci.* **1968**, *33*, 223-226.
13. Tunick, M.H.; Nolan, E.J. In *Physical Chemistry of Food Processes;* Baianu, I.C., Ed.; Van Nostrand Reinhold: New York, NY, 1992; pp 273-297.
14. Diehl, K.C.; Hamann, D.D.; Whitfield, J.K. *J. Text. Stud.* **1979**, *10*, 371-400.
15. Tunick, M.H.; Van Hekken, D.L. *J. Dairy Sci.* **2002**, *85*, 2743-2749.
16. Truong, V.D.; Daubert, C.R. *J. Food Sci.* **2001**, *66*, 716-721.
17. Tunick, M.H.; Van Hekken, D.L. *J. Text. Stud.* **2003**, *34*, 219-229.
18. Dryer, J. *Dairy Foods* **1994**, (6), 20.
19. Malin, E.L.; Brown, E.M. In *Chemistry of Structure-Function Relationships in Cheese*; Advances in Experimental Medicine and Biology, No. 367; Malin, E.L., Tunick, M.H., Eds.; Plenum Press: New York, NY, 1995; pp 303-310.
20. Tunick, M.H.; Mackey, K.L.; Shieh, J.J.; Smith, P.W.; Cooke, P.; Malin, E.L. *Int. Dairy J.* **1993**, *3*, 649-662
21. Tunick, M.H.; Malin, E.L.; Smith, P.W.; Holsinger, V.H. *Cult. Dairy Prod. J.* **1995**, (5), 6-9.

Flavor and Texture

Chapter 12

How Lipids Influence Flavor Perception

Kris B. de Roos

Givaudan Nederland B.V., P.O. Box 414, 3770 AK Barneveld,
The Netherlands

Lipids affect flavor perception by their effect on aroma, taste and mouthfeel. Aroma and taste are influenced via the effect that lipids have on phase partitioning and mass transport, whereas mouthfeel is affected by the direct effect of the lipids on texture. Results of in vivo studies indicate that flavor release during consumption is strongly diffusion controlled. Therefore, the release is sensitive for factors that affect mass transport in the product phase, such as the content, consistency and particle size of the lipids in emulsions. The effect of lipids on the overall flavor is complex due to the interactions between aroma, taste and mouthfeel.

Introduction

Lipids are hydrophobic compounds with low solubility in water. When mixed with water, they form a separate phase, which has a totally different affinity to flavor compounds than the aqueous phase. Therefore, the presence of lipids in foods has often a strong effect on flavor release and perception.

To understand how lipids influence flavor perception it is necessary to know the sensations that contribute to flavor. In this review we will use the broad definition of flavor that says that flavor is the result of the combined effects of odor, taste and mouthfeel ([1]):

- **Odor** or **aroma** is the result of the stimulation of receptors in the nose by volatile chemicals. To be perceptible, aroma compounds must be volatile to allow transport via the air to the olfactory epithelium in the nose.

- **Taste** is the result of stimulation of receptors in the mouth by volatile or non-volatile chemicals dissolved in the saliva. This definition of taste does not only include real taste sensations such as sweetness and acidity but also trigeminal and other sensations such as astringency, pungency and soapiness.
- **Mouthfeel** results from tactile sensations. These sensations can be perceived by touch and allows perception of differences in texture, structure and temperature. Mouthfeel is the result of *physical* stimulation of receptors in the mouth in contrast with taste, which is the result of *chemical* stimulation.

Lipids can influence flavor perception either directly by their effect on flavor release and texture or indirectly via their flavor precursor properties and their effect on the flavor stability (*2*). This review will be restricted to the first category of direct effects.

Effect of lipids on aroma release and perception

The physico-chemical parameters that control the aroma release from products are the air-product partition coefficient P_{ap} and the mass transfer coefficient k. Both parameters are strongly influenced by the presence of the lipids in a product (*2, 3*).

Effect of lipids on phase partitioning

An aroma compound, when allowed to equilibrate between a product and air, distributes over the two phases according to the air-product partition coefficient P_{ap}, which is defined as:

$$P_{ap} = C_a / C_p \qquad (1)$$

where C_a and C_p are the concentrations (g/cm^3) of the aroma compound in the air and product phase, respectively.

The air-product partition coefficient, which is a measure for the volatility of a compound in a product, is strongly dependent on the product composition. The differences in the volatility of aroma compounds in different products reflect their affinity to these products. The difference in volatility in water and vegetable oil, as expressed by P_{aw} and P_{ao}, is a measure for the lipophilicity (hydrophobicity) of an aroma compound and is most conveniently expressed by the oil-water partition coefficient P_{ow}:

$$P_{aw}/P_{ao} = (C_a/C_w)/(C_a/C_o) = C_o/C_w = P_{ow} \qquad (2)$$

An oil useful as reference is olive oil, which is the moderately unsaturated. However, traditionally octanol is often being used as the reference medium. Although octanol-water partition coefficients satisfactory predict the partitioning between water and amphiphilic lipids, such as phospholipids (4) and mono- and diglycerides, the prediction is much less satisfactory with neutral lipids such as triglycerides (5, 6) and waxes (like likes like).

The volatility in emulsions can be calculated from the volume fractions f_o and f_w of oil and water, and the partition coefficients P_{oa} and P_{wa}:

$$P_{ap} = C_a/C_p = \frac{C_a}{f_o C_o + f_w C_w} = \frac{1}{f_o P_{oa} + f_w P_{wa}} \qquad (3)$$

Figure 1 shows how lipids affect the volatility (P_{ap}) of aroma compounds of different hydrophobicity (limonene is most and diacetyl least hydrophobic).

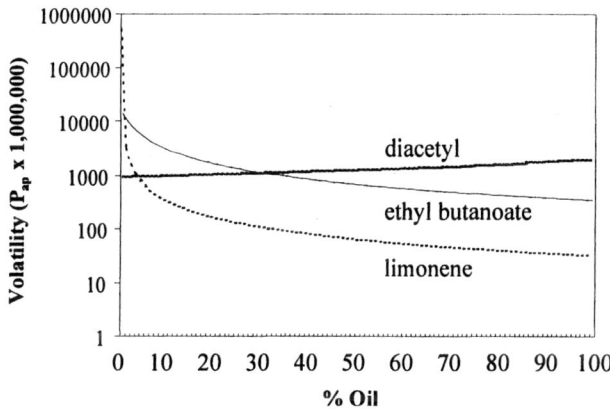

Figure 1. Volatility in emulsions as a function of the oil volume fraction (according to eq 3)

The exact relationships of equations 2 and 3 hold only if the two phases are not soluble in each other. With octanol and low-molecular-weight triglycerides, such as triacetin and tributyrin, both phases are mixtures of water and lipid in which the aroma volatility is different from that in the pure phases. In such systems one has to distinguish between two types of interactions: interactions of the aroma compounds with dissolved lipids and interactions with aggregates such as oils droplets and micelles. The first type of interactions is weak compared to the interactions of the second type and can often be neglected if they occur at the same time (*7*).

The oil droplet size in emulsions does not significantly affect the phase partitioning, while the fatty acid chain length (*8-11*) and the degree of unsaturation of the lipids have only a minor effect provided that they do not change the solid fat content (*5, 9, 12*). The effects of chain length and unsaturation can be predicted on the basis of the "like likes like" principle. So, short-chain esters have highest affinity for triglycerides with short fatty acid chain length, while the more hydrophobic aroma molecules have highest affinity to the hydrophobic high-molecular-weight triglycerides (*5*).

A factor that has a major effect on phase partitioning is crystallization. Dissolved solutes are excluded from the crystal lattice, which results in higher solute concentrations in the remaining liquid part of the crystallizing phase and in the phases that are in equilibrium with it. Ice formation results in a quantitative exclusion of other solutes from the crystal lattice (*13*) and recent work on solid fats suggest that fat crystallization might have the same effect (*9, 14*). However, the results are not consistent (*5, 12*). This might be due to the different degrees of crystallinity of the fats. Amorphous and less rigid polymorphous crystalline areas in the fats can easily incorporate lipophilic aroma compounds.

Amphiphilic lipids differ from neutral lipids in their ability to adsorb to hydrophilic particles. The result is that these particles become more hydrophobic and that the adsorption of hydrophobic aroma compounds increases (*15*). This is why depulping of juices can lead to high aroma loss. Absorption of aroma compounds in cellular structures might here play a role as well. Lipids present in yeast or plant cells can absorb high proportions of lipophilic aroma compounds (*16*).

Temperature affects the volatility of aroma compounds in water and lipids to different extents (*17*). This means that the partitioning between lipids and water is also affected to a different extent (*18-20*). Or in other words, the relative affinity of aroma compounds to lipids changes with the temperature.

Effect of lipids on mass transport

When air is sweeping across a food and dilutes the headspace concentrations, mass transport from product to air will take place in an effort to restore the phase equilibria. This results in concentration gradients in the product and vapor phase as depicted in Figure 2. The degree of non-equilibrium, represented by the concentration gradients ΔC_w and ΔC_a, is the driving force for mass transport, while the resistance to that transport is given by $R = 1/k$. So, the mass flux J in either phase is then given by (*3*):

$$J = \Delta C / R = k(C^i - C) \tag{4}$$

where J is expressed in g/cm^{2}s, k is the mass transfer coefficient (cm/s) and C and C^i are the aroma compound concentrations (g/cm^3) in the bulk phase and at the interface.

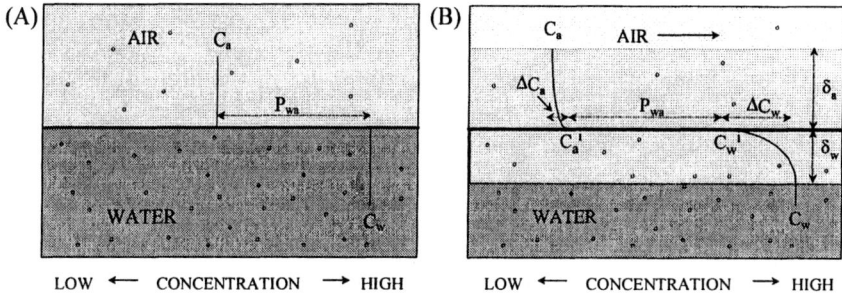

Figure 2. Concentration gradients of aroma compounds in water and air under static and dynamic conditions

The concentration gradients generated at the product surface increase with the rate of aroma volatilization and the resistance to mass transfer. The higher the resistance, the more difficult it will be to replenish depleted concentrations at the product surface.

Since it is the aroma compound concentrations at the product surface that determine the maximum concentrations in the air, it is clear that under dynamic conditions the maximum headspace concentrations predicted by equations 1 and 3 will almost never be achieved. At very high flow rates over the product surface and/or very low mass transfer rates in the product, the aroma extraction from the product surface is exhaustive and $C_p^i \to 0$. The mass flux J_p in the product is then a function of only the concentration C_p in the bulk phase and the mass transfer coefficient k_p:

$$J_P = -k_P C_P \tag{5}$$

The value of the mass transport coefficient k is strongly dependent on the diffusion mechanism. In a stagnant phase the only mechanism of mass transport is the molecular or static diffusion, which is caused by the random movement of the molecules. The rate of molecular diffusion is determined by the diffusion coefficient D, which varies with the viscosity of the medium according to the Stokes-Einstein equation:

$$D = k_B T / 6\eta\pi r \qquad (6)$$

where k_B = Boltzman's constant, T = temperature, η = dynamic viscosity, and r = radius of the molecule. In monophasic systems, differences between diffusion coefficients are small, since the radii r of aroma molecules do not vary much.

More variation of the diffusion coefficients is observed in emulsions. This is due to the different diffusion rates in aqueous and lipid phases and the unequal distribution of the flavor compounds over these phases (21). Since the diffusion constants in lipids are lower than those in water (about 10^{-10} m^2/s in oil versus 10^{-9} m^2/s in water), the static diffusion in emulsions decreases with the lipophilicity of the flavor compounds and the lipid content of the product.

In a dynamic phase the eddy or convective diffusion is the most important mechanism of mass transport (3). Eddy diffusion carries elements or eddies of the product phase from one location to another and is completely independent of flavor compound type. The diffusion increases with the kinetic energy put in the system and decreases with its viscosity. Since eddy diffusion transports lipids and flavor compounds at the same rate, lipids will not affect the mass transport if the partitioning between lipid and water phase is instantaneous as is often assumed (22, 23). This would mean that a completely eddy diffusion controlled release is related to the total concentration of the aroma molecules in the emulsion. This is in contrast with the release under equilibrium conditions, which is related to the concentration of molecules in the aqueous phase.

Mass transport during retronasal aroma perception

Linforth **et al**. (10) have demonstrated that the aroma release during drinking is strongly kinetically controlled. Comparison of the release under equilibrium conditions with that during drinking showed that the release rates during drinking were much more similar. This was due to a strong reduction of the release of the most volatile compounds, which is typical for a strongly kinetically controlled release. The concentrations in the breath from the nose were lower than those from the mouth but the trend in the relative release rates was the same demonstrating that in both cases the release was strongly kinetically controlled. The uniformity of the *in vivo* release indicates that the (diluted) product in the mouth has been exhaustively extracted at its surface as a result of very high airflow rates passing over it (24, 25).

The effect of lipids on the retronasal aroma release is also much smaller than that on the equilibrium release (25-28). Figure 3 shows the difference between the release of hydrophilic and lipophilic volatiles from a turbulent emulsion. For water-soluble compounds the only mechanism of mass transport to the surface is that from the bulk phase (Figure 3A), whereas for lipophilic compounds there is an additional mechanism consisting of a mass transport from the lipid particles in boundary layer δ_w to the surface of the emulsion (Figure 3B). At equal concentrations in the aqueous bulk phase (equal equilibrium headspace concentrations), this results in an enhanced release of the lipophilic aroma compounds.

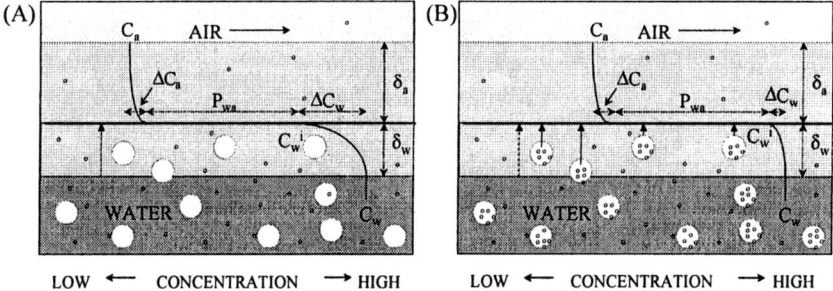

Figure 3. Concentration gradients in water and emulsions for a hydrophilic (A) and hydrophobic aroma compound (B) with equal volatility in water

Under the strongly diffusion controlled conditions during consumption, one may expect that the consistency and particle size of the lipid phase in emulsions have also an effect on the aroma release. With regard to the oil droplet size, this is indeed what has been observed: a smaller oil droplet size results in higher release rates *(10, 29, 30)*. The same effect of oil droplet size on the perceived intensity has been observed during smelling *(11)*.

Surprisingly, model mouth studies showed a decrease of the aroma release with decreasing oil droplet size *(22)*. Apparently, under the less kinetically controlled conditions of the model mouth the positive effect of smaller oil droplet size is more than nullified by the negative effect of the higher emulsion viscosity. The result is in agreement with theoretical models that assume equilibrium between lipid and aqueous phase during the dynamic flavor release *(23)*. From the difference between the results of the *in vivo* and model mouth studies it may be concluded that during the strongly kinetically controlled release in the mouth no equilibrium exists between lipid and aqueous phase.

In contrast with the previous study, Charles *et al.* *(31)* have found that an increase of the oil droplet size in emulsions selectively reduces the dynamic release of the hydrophobic aroma compounds. At the same time the release of the hydrophilic compounds increased, which was again attributed to the low viscosity at high oil droplet size. The decrease of the release of the hydrophobic compounds was assumed to be caused by the "shell of immobilized water", created around the oil droplets by the adsorbed hydrophilic emulsifiers. This shell of immobilized water, which increases the diffusion pathway from oil to water, increased with growing particle size (decreasing oil-water interfacial surface area). Experiments with coacervate microcapsules have shown that a shell of immobilized water around an oil droplet can indeed lead to a major reduction of the release of the hydrophobic aroma compounds (*32, 33*).

Mass transport during orthonasal aroma perception

The orthonasal aroma (smell) of liquid and semi-solid products is less influenced by the kinetics of the flavor release than the retronasal aroma (*3, 26, 27*). This can be concluded from the effect of fat on the smell, which is more according to the aroma volatility than the effect of fat on the retronasal aroma (Figure 4). One of the consequences of the difference between the orthonasal and aroma retronasal perception is that low-fat products have in general a stronger smell than their full-fat analogues if the retronasal aroma is the same.

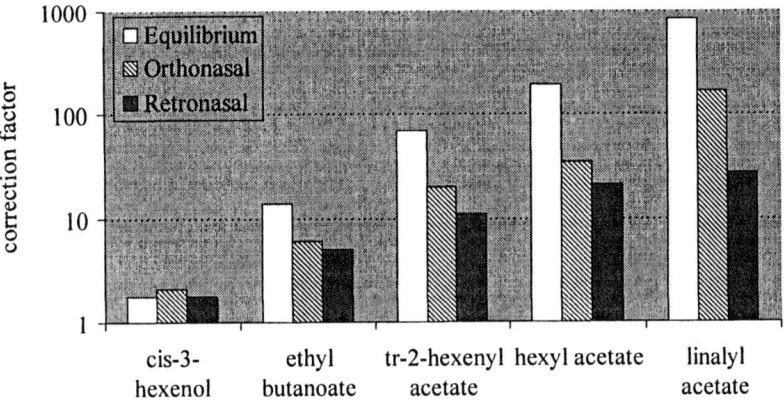

Figure 4. Correction factors for same flavor intensity in cream (33% fat) as in a soft drink under equilibrium and non-equilibrium conditions (3)

With solid products, such as biscuits or hard candies, the situation is different. In such products, the diffusion of the aroma compounds through the hydrophilic phase is strongly hindered, if possible at all (*33*). The lipid phase, on the other hand, is a poor aroma barrier. Therefore, the orthonasal aroma of solid products is mainly coming from the lipid phase. Rapid release of the flavor compounds immobilized in the hydrophilic phase is only possible when during consumption the hydrophilic phase is hydrated or dissolved. In general, dry products with no or low fat have lower orthonasal than retronasal impact.

Effect of lipids on time-intensity profile of aroma release

Rebalancing of flavorings to provide the same maximum aroma intensity I_{max} in low and full-fat products does not always result in the same aroma perception. One of the major reasons is that lipids influence the temporal profile of the release of each flavor compound to a different extent. The duration of the release increases with the hydrophobicity of the flavor compound and the lipid content of the product (*28, 30, 32, 34*). In full-fat products this results in a change of aroma profile with time (Figure 5) thus generating a flavor sensation that is perceived as rich. A serious flavor defect of low fat foods is the quick disappearance of the flavor in the mouth and the lack of richness.

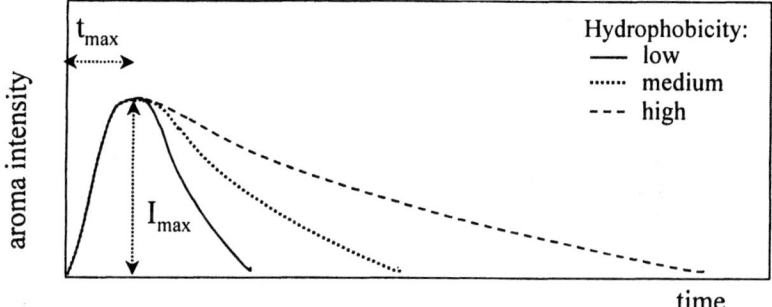

Figure 5. Schematic picture of the effect of flavor hydrophobicity on the temporal profile of the release from iso-viscous emulsions

To overcome the flavor defects of low-fat foods, controlled delivery systems have been developed to prolong the rate of the lipophilic flavor release. One of these methods consists of adding oil containing particles to a product, which absorb the lipophilic flavor compounds (*34-37*). The result is a delayed release of the absorbed lipophilic compounds during consumption due to the increased effective path length of the diffusion from the oil droplets to the no- or low-fat environment (*34-36*). The release of the water-soluble flavor compounds is not affected because they remain outside the gel particles. Another method to prolong the release from low fat products consists of encapsulating the aroma compounds in fat particles that release the aroma compounds when melting in the mouth during consumption (*38*).

Lipids can sometimes also affect the time to maximum aroma intensity (t_{max}) during consumption. Malone *et al.* (*34*) found that t_{max} increases with the oil content when measuring the perceived aroma intensity by means of sensory methods during consumption of iso-viscous oil-in-water emulsions. However, when measuring the aroma compound concentrations in the breath from the nose, the investigators found that the maximum intensity was achieved independent of the oil content. This might indicate that the time to maximum intensity is related to the rate of receptor saturation in the nose (the release rates decrease with the oil content in the product).

Whereas emulsions of liquid oil do often not show an increase of t_{max} with increase of oil content (*27, 32, 34*), those with solid fat do (*2, 29, 39*). The more hydrophobic an aroma compound and the higher the fat content of the product, the longer is t_{max}. The flavor release seems to be related to fat melting behavior, the release being postponed till the fat is melting (*40*). This probably explains the delayed perception of the lipophilic linalyl acetate in butter, which is clearly separated from the immediate perception of the water-soluble diacetyl (*2*).

Effect of lipids on taste perception

In general, taste is much less influenced by lipids than aroma due to the low hydrophobicity of most taste compounds. Exceptions are glycyrrhizin, capsaicin, menthol (as cooling agent) and some other hydrophobic taste compounds. The concentrations of these compounds need to be increased with increasing lipid content to achieve the same flux to the tongue as in the absence of lipids. Compounds with both taste and odor properties, such as menthol, might require different corrections for same taste and aroma impact due the differences in the resistance to mass transfer from oil to saliva and saliva to air.

Lipids can also affect the perceived intensity of water-soluble taste compounds. In general, an increase of the lipid content results in an increase of taste intensity because the taste compounds become more concentrated in a smaller volume of aqueous phase. Although a high lipid content has also negative effects on the perceived taste intensity (due to mouth coating, decrease of aqueous volume and smaller contact surface area between aqueous phase and mouth) the overall result is that the intensity of water-soluble taste compounds increases with an increase of the lipid content (34 and references cited).

Effect of lipids on mouthfeel

The poor texture of low fat products has been one of the major reasons of the low consumer acceptance. Therefore, major emphasis has been placed in the past on texture and water binding to provide the rich creamy mouthfeel lost when fat was removed. With the wide range of hydrocolloids, texture modifiers and fat replacers currently available, this problem is now often satisfactorily to solve. Whereas the standard fat replacers provide only mouthfeel, the gel particles with encapsulated oil droplets, mentioned above (34), provide also the for full-fat foods characteristic extended release of lipophilic compounds.

Effect of lipids on total flavor

The average consumer has problems with distinguishing between aroma, taste and mouthfeel; he or she perceives flavor as a single sensation. This is one of the reasons why the effect of lipids on total flavor is perceived as complex. Another reason is that each of these sensations is influenced by one or both of the other sensations (Figure 6).

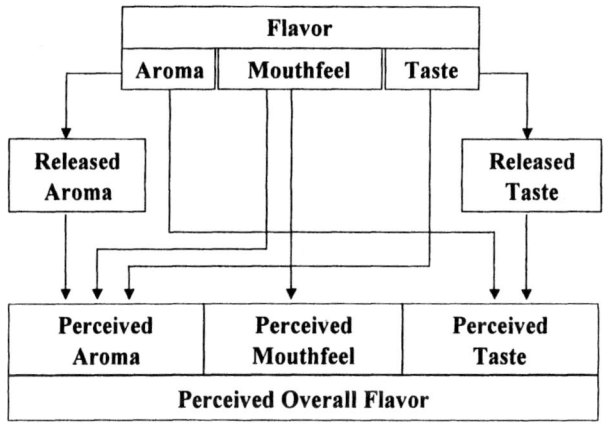

Figure 6. Interactions between aroma, taste and mouthfeel and their effect on overall flavor perception

Several examples are known that show that the perceived aroma is not only due to the concentrations of volatile compounds in the nose but also to the simultaneous taste and mouthfeel sensations (41-48). Taste compounds have been found to enhance the perception of aroma compounds (44-47) and vice versa (49). If either taste or aroma is missing, the flavor is not perceived as complete. This holds in particular, with regard to taste; loss of sweetness often gives the impression that the total flavor has been lost (42, 48).

Despite the widespread interest in total fat reduction, many people still prefer to eat the full-fat products. This gap between actual eating pattern and the marked concern with healthy food is assumed to arise primarily from the compromise in the flavor of low fat foods. Although there are indications that clean fat, independent of its viscosity and flavor (50), increases already the palatability of foods, it would be interesting to know which lipid induced changes in the flavor release are contributing to a high consumer preference. Although in recent years already considerable progress has already been made in improving the flavor of reduced-fat foods, continuing improvement is desirable to further increase the acceptance of reduced-fat foods by the consumer.

References

1. Ney K.H. *Gordian* **1988**, *88 (1)*, 19.
2. De Roos, K.B. *Food Technol.* **1997**, *51 (1)*, 60-62.
3. De Roos, K.B. In: *Flavor Release,* Roberts, D.D.; Taylor, A.J., Eds.; American Chemical Society: Washington, D.C., 2000, pp. 126-141.
4. Chen, N.; Zhang, Y.; Terabe, S.; Nakagawa, T. *J. Chromatogr.* **1994**, *678*, 327-332.
5. Rabe, S.; Krings, U.; Zorn, H.; Berger, R.G. *Lipids* **2003**, *38*, 1075-1084.
6. Rabe, S.; Krings, U.; Berger, R.G. *Food Chem.* **2004**, *38*, 117-125.
7. Hussam, A.; Basu, S.C.; Hixon, M.; Olume, Z. *Anal. Chem.* **1995**, *67*, 1459-1464.
8. Carey, M.; Asquith, T.; Linforth, R.S.T.; Taylor, A.J. *J. Agric Food Chem.* **2002**, *50*, 1985-1990.
9. Roberts, D.D.; Pollien, P.; Watzke, B. *J. Agric. Food Chem.* **2003**, *51*, 189-195.
10. Linforth, R.; Martin, F.; Carey, M.; Davidson, J.; Taylor, A.J. *J. Agric. Food Chem.* **2002**, *50*, 1111-1117.
11. Miettinen, S.-M.; Tuorila, H.; Piironen, V.; Vehkalathi, K.; Hyvönen, L. *J. Agric. Food Chem.* **2002**, *50*, 4232-4239.
12. Rabe, S.; Krings, U.; Berger, R.G. *J. Sci. Food Agric.* **2003**, *83*, 1124-1133.
13. Thijssen, H.A.C.; Van der Malen, B.G.M., *Eur. Pat. Appl.* EP 15042, 1980.
14. Roudnitzky, N.; Irl, H.; Roudaut, G.; Guichard, E. In: *Flavour Research at the Dawn of the Twenty-first Century*; Le Quéré, J.L.; Étiévant, P.X., Eds.; Editions Tec & Doc: Paris, 2003, pp 136-139.
15. Brat, P.; Rega, B.; Alter, P.; Reynes, M.; Brillouet, J.M. *J. Agric. Food Chem.* **2003**, *51*, 3442-3447.
16. Bishop, J.R.P.; Nelson, G.; Lamb, J. *J. Microencapsulation* **1998**, *15*, 761-773.
17. Hall, G.; Anderson, J. *Lebensm. Wiss. u. Technol.* **1983**, *16*, 362.
18. Kertes, A.S.; King, C.J. *Chem. Rev.* **1987**, *87*, 687-710.
19. Kinkel, J.F.M. ; Tomlinson, E.; Smit, P. *Int. J. Pharm.* **1981**, *9*, 121-136.
20. Leo, A.; Hansch, C.; Elkins, D. *Chem. Rev.* **1971**, *71*, 525-616.
21. Overbosch, P.; Afterof, W.G.M.; Haring, P.G.M. *Food Rev. Int.* **1991**, *7*, 137-184.

22. Van Ruth, S.M.; King, C.; Giannouli, P. *J. Agric. Food Chem.* **2002**, *50*, 2365-2371.
23. Harrison M.; Hills, B.P.; Bakker, J.; Clothier, T. *J. Food Sci.* **1997**, *62*, 653-658, 664.
24. Espinosa-Diaz, M.; De Roos, K.B.; Antenucci, R.N. In: *Flavour Research at the Dawn of the Twenty-first Century*; Le Quéré, J.L.; Étiévant, P.X., Eds.; Editions Tec & Doc: Paris, 2003, pp. 83-86.
25. De Roos, K.B.; Wolswinkel, C. In: *Trends in Flavour Research*; Maarse, H.; Van der Heij, D.G., Eds.; Elsevier Science B.V.: Amsterdam, 1994, pp. 15-32.
26. Roberts, D.D.; Pollien, P.; Antille, N.; Lindinger, C.; Yeretzian, C.B. *J. Agric. Food Chem.* **2003**, *51*, 3636-3642.
27. Miettinen, S.-M.; Hyvönen, L.; Tuorila, H. *J. Agric. Food Chem.* **2003**, *51*, 5437-5443.
28. Doyen, K.; Carey, M.; Linforth, R.S.T.; Marin, M.; Taylor, A.J. *J. Agric. Food Chem.* **2001**, *49*, 804-810.
29. Brauss, M.S.; Linforth, R.S.T.; Cayeux, I.; Harvey, B.; Taylor, A.J. *J. Agric. Food Chem.* **1999**, *47*, 2055-2059.
30. Carey, M.; Linforth R.; Taylor, A. In: *Flavor Research at the Dawn of the Twentieth-first Century*; Le Quéré, J.L.; Étiévant, P.X., Eds.; Editions Tec & Doc: Paris, 2003, pp. 212-215.
31. Charles, M.; Rosselin, V.; Beck, L.; Sauvageot, F.; Guichard, E. *J. Agric. Food Chem.* **2000**, *48*, 1810-1816.
32. Malone, M.E.; Appelqvist, I.A.M.; Norton, I.T. *Food Hydrocolloids* **2003**, *17*, 775-784.
33. De Roos, K.B. *Int. Dairy J.* **2003**, *13*, 593-605.
34. Malone, M.E.; Appelqvist, I.A.M.; Goff, T.C.; Homan, J.E.; Wilkins, P.G. In: *Flavor Release*; Roberts, D.D.; Taylor, A.J., Eds.; American Chemical Society: Washington, D.C., 2000, pp. 212-227.
35. Malone, M.E.; Appelqvist, I.T. *J. Controlled Rel.* **2003**, *90*, 227-241.
36. Lian, G. In: *Flavor Release*; Roberts, D.D.; Taylor, A.J., Eds.; American Chemical Society: Washington, D.C., 2000, pp. 201-211.
37. Bouwmeesters, J.F.G.; De Roos, K.B. *WO Patent* 9815192, 1998.
38. Graf, E.; De Roos, K.B. In: *Flavor-Food Interactions*; McGorrin, R.J.; Leland, J.V., Eds.; American Chemical Society: Washington, D.C., 1996, pp. 24-35.
39. Ingham, K.E.; Taylor, A.J.; Chevance, F.F.V.; Farmer, L.J. In: *Flavor Science. Recent developments*; Taylor, A.J.; Mottram, D.S., Eds.; The Royal Society of Chemistry: Cambridge, UK, 1996, pp. 386-391.
40. Andreasen, L.V.; Horndrup, B.; Marcussen, J. In: *Flavour Research at the Dawn of the Twenty-first Century*; Le Quéré, J.L.; Étiévant, P.X., Eds.; Editions Tec & Doc: Paris, 2003, pp. 200-203.

41. King, B.M.; Arents, P.; Derks, E.P.P.A.; Duineveld, C.A.A.; Boelrijk, A.E.M.; Burgering, M.J.M. In: *Flavour Research at the Dawn of the Twentieth-first Century*; Le Quéré, J.L.; Étiévant, P.X., Eds.; Lavoisier, Cachan: France, 2002, pp. 164-169.
42. Taylor, A.; Hollowood, T.; Davidson, J.; Cook, D.; Linforth, R. In: *Flavour Research at the Dawn of the Twentieth-first Century*; Le Quéré, J.L.; Étiévant, P.X., Eds.; Editions Tec & Doc: Paris, 2003, pp. 194-199.
43. Weel, K.G.C.; Boelrijk, E.M.; Alting, A.C.; Van Mil, P.J.J.M.; Burger, J.J.; Gruppen, H.; Voragen, A.G.J.; Smit, G. *J. Agric. Food Chem.* **2002**, *50*, 5149-5155.
44. Duizer, L.M.; Bloom K.; Findlay, C.J. *J. Food Sci.* **1996**, *61*, 636-638.
45. Noble A.C. *Trends Food Sci. Technol.* **1996**, *7*, 439-444.
46. Noble, A..C.; Kuo, Y.; Pangborn, R.M., *Int. J. Food Sci. Technol.* **1993**, *28*, 127-137.
47. Valdés, R.M.; Hinreiner, E.H.; Simone, M.J. *Food Technol.* **1956**, *10 (6)*, 282-285.
48. Davidson, J.M.; Linforth, R.S.; Hollowood, T.A.; Taylor, A.J. *J. Agric. Food Chem.* **1999**, *47*, 4336-4340.
49. Frank R.A.; Byram, J. *Chem. Senses* **1988**, *13*, 445-455.
50. De Araujo, I.V.; Rolls, T. *J. Neurosci.* **2004**, *24*, 3086-3093.

Chapter 13

Release of Flavor from Emulsions under Dynamic Sampling Conditions

R. S. T. Linforth and A. J. Taylor

School of Biosciences, Division of Food Sciences, University of Nottingham, Sutton Bonington Campus, Loughborough, Leics LE12 5RD, United Kingdom

In vivo flavor release studies were performed to determine the effect of lipid on volatile delivery during eating. Lipids created a reservoir for stable delivery of flavor molecules throughout the eating time course. The delivery of flavor molecules from lipid containing systems was greater than expected on the basis of static headspace analysis. This was attributed mainly to changes in mass transfer. Mass transfer was dependent on the product/air partition coefficient. Decreasing the partition coefficient, through the solubilization of lipophilic compounds in the oil phase, increased mass transfer. This effectively reduced the impact of a lower partition coefficient on overall flavor delivery. Whereas lower lipid content systems (associated with higher product/air partition coefficients) had poor mass transfer characteristics, and delivered flavor inefficiently under dynamic conditions.

Volatile flavor compounds interact to different extents with lipids in foodstuffs, dependent on their physicochemical properties. Consequently, low fat foods can deliver incorrectly balanced flavor profiles (relative to higher fat content formulations) as the partitioning of flavor within the food matrix is affected (*1*). The lower fat product is also considered to give a more intense release of flavor early during consumption and have less persistence than the high fat content equivalent (*1,2*). Why is this? The change in the balance of flavor compounds could be explained by simple partitioning of molecules into the lipid decreasing their availability. However, the fact that lipid is associated with less flavor early in the eating process but, can enhance persistence (i.e. results in sustained flavor delivery) requires a more complex hypothesis.

One theory (*3*), is that during consumption, the dilution of the lipid within the food results in an increase in the product/air partition coefficient hence altering delivery as eating continues. This is based on the observation that the partition of molecules between the product and air, depends on both their affinity for lipid relative to water, and the size of the oil and water fractions (*4*). Any change in the oil fraction by dilution with saliva may affect flavor partitioning.

In addition there is the potential influence of lipid on mass transfer to consider. Hills and Harrison (*5*), used the effect of lipid on the viscosity of the system to modify mass transfer in their model, which described the effect of emulsions on flavor release. However, lipids can have profound effects on flavor delivery (*6*) even at low concentrations which cause an insignificant change in viscosity. Doyen and co-workers (*6*), used the direct effect of lipid on the product/air partition coefficient as the key component of their mass transfer equation, based on earlier studies which had shown the influence of the partition coefficient on mass transfer (*7*).

It is likely, that both mass transfer and changes in partition (due to dilution) will influence release during consumption, the extent of which will depend on the composition of the bolus and the duration of the eating process. The influence of these two factors on flavor delivery will be examined in this paper.

Materials and Methods

Gas Phase Analysis

The gas phase was directly sampled (at flow rates ranging from 4 to 40mL/min) into an Atmospheric Pressure Chemical Ionization source of a

Platform LCZ mass spectrometer (MS Nose™, Micromass, Manchester UK). The transfer line (1m x 0.53mm ID, deactivated fused silica) was maintained at 125°C to avoid the condensation of water or flavor molecules. Once in the source the molecules were ionized by a 4kV corona discharge, which resulted in their protonation (positive ion mode). These ions were monitored in selected ion mode with a dwell time of 0.02s, which was sufficiently short to allow the rapid changes in volatile concentration in the breath to be followed.

Sample Preparation and Consumption

Emulsions were prepared by dispersing, gum arabic (96g/L), citric acid (10g/L) and potassium sorbate (25mg/L) in water using a high shear blender (Silverson Machines Ltd, Chesham, UK) for 10min before addition of the lipid (96g/L), and blending for a further 25min. The solution was then passed through a homogenizer (APV, Crawley, UK) three times at 4500psi. The emulsion was stored at 4°C and diluted to the appropriate concentrations before flavor addition (1 to 100mg/L). For the yogurt samples, flavor was added to the mixed, hydrated, pasteurized yoghurt ingredients prior to fermentation.

Aliquots of these samples (typically 15mL) were placed in the mouth and swallowed whilst the breath flavor concentration was monitored. Headspace samples were prepared by equilibrating the solutions in stoppered 100mL flasks for 1h at 22°C, the stopper was then removed to allow the transfer line of the MS Nose to be introduced into the flask for gas phase sampling.

Results and Discussion

Flavor Delivery From Yogurt

Lipid reduced the intensity of flavor delivery: yoghurt containing either 3.5% or 10% lipid, gave substantially lower maximum anethole breath volatile concentrations than a 0.2% lipid content sample (Figure 1). In addition the area under the release curve (representative of the total amount of flavor delivered) was also lower for the 3.5% and 10% lipid content samples compared with the 0.2% sample. This may have been due to the loss of the flavor molecules as they partitioned into the lipid and became effectively lost. The effects of small amounts of lipid were substantial, with lesser changes in flavor delivery occurring as the lipid concentration was increased further. Consequently, the release profiles for the 10% and 3.5% lipid samples were broadly similar, and much lower in intensity than the 0.2% lipid sample. This should mean that reduced fat foods are feasible without substantially altering flavor delivery. However, the reduction in lipid content can also affect the texture and mouthfeel of the product, which can have a significant impact on its overall perception.

The effect of lipid was not however just one of altering the magnitude of release, it also influenced the temporal dimension of release (Figure 2). The higher lipid content yogurts were associated with the most persistent (intensity, relative to maximum intensity) breath volatile contents. This demonstrated that the flavor molecules were not lost and bound in the lipid, but were available as a reservoir for further delivery. This could clearly be due to either changes in mass transfer, or the influence of dilution on partition. To investigate this further we needed to use a simpler sampling regime.

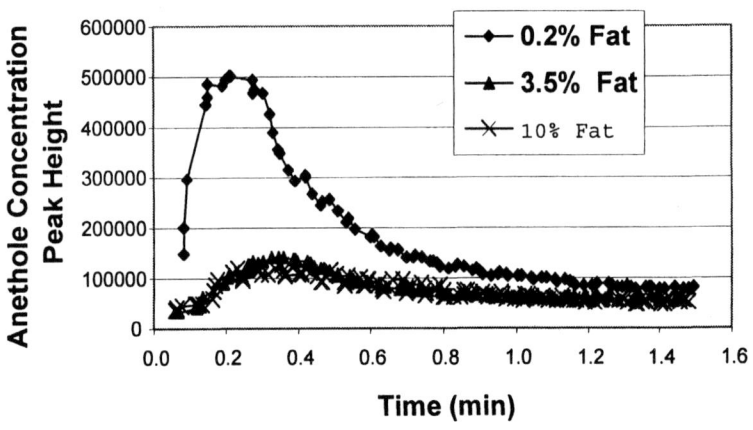

Figure 1. The release of anethole (400mg/Kg) from yogurt with different fat contents during consumption (each curve is the average of 5 replicates). (Reproduced from reference 2. Copyright 1999 ACS)

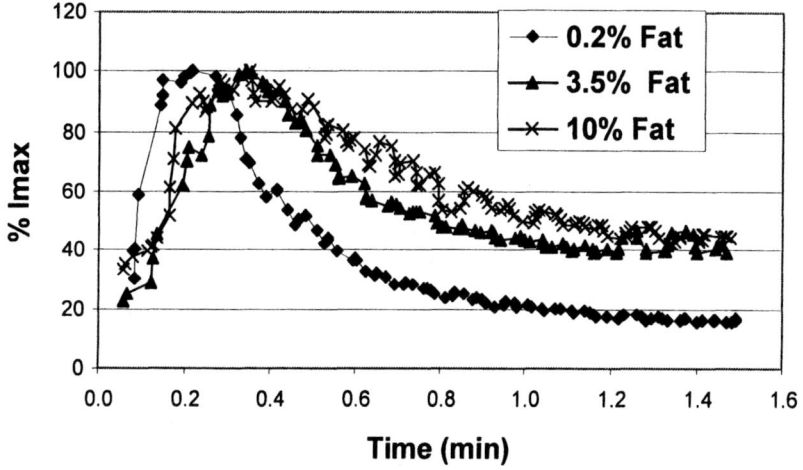

Figure 2. The release of anethole from yogurt with different fat contents during consumption after normalization of the intensity data as a % of the maximum (Imax). (Reproduced from reference 2. Copyright 1999 ACS)

Flavor Delivery From Simple Solutions: Profile Shape

Consumption of flavors in either water or an emulsion can be achieved with one single swallowing action, resulting in a simplified eating action which may be easier to interpret. The persistence of volatile compounds in the breath can be followed, as the intensity of volatile in the breath after swallowing relative to the maximum breath concentration. This will arise from both residual sample in the throat, and from volatiles absorbing and subsequently desorbing from the surface of the nasal epithelium, the wash – in wash – out principle (8). Volatiles residing in the mouth are unlikely to pass into the breath without further swallowing or substantial mouth movement (9).

Typical breath volatile profiles for the aqueous and lipid containing samples are shown in Figure 3. The profiles were clearly different, with the lipid containing sample showing the greatest persistence. The enhanced persistence, could have been induced by the dilution of the sample residue left after swallowing, changing the product/air partition coefficient and hence delivery. If this were the case (and dilution did occur), we should be able to see a greater decrease in the signal for the aqueous sample (relative to that of the emulsion) as this would be unable to alter its partition upon dilution.

However, the rate of decline of the latter parts of the release curves were similar in both cases which implies that there were no substantial amounts of dilution occurring after the first exhalation. Indeed, if substantial dilution had occurred the profile for ethyl octanoate in water (lower right) should have shown a greater rate of decline. However, it only fell by 60% over 1min, which would indicate that very little dilution could actually have occurred; not the 10 to 20 fold that would be necessary to make significant change the product/air partition coefficient (6). It is more likely that the changes in ethyl octanoate observed were due to losses into the gas phase.

The phase of the profiles detailed on the right in Figure 3, would be comparable to the latter part of the release curves in Figure 1 (after the main mouth movement and swallowing phase, i.e. after approximately 20s). Here too there was a slow decline in the breath anethole concentration for the low lipid (0.2%) yogurt after 0.5min, consistent with minimal dilution occurring.

The only point where a major difference in the shape of the emulsion and water sample profiles occurred was between the first and second exhalations (Figure 3), implying that dilution may have occurred here. However, it would have been unlikely for substantial dilution of the bolus (15mL) to take place during the initial swallowing action due to the low amounts of saliva typically present in mouth.

It is possible to argue however, that flavor delivery was dependent on a small part of the bolus (rather than the main bulk of it), and that this important fraction of the sample may have been significantly diluted. If such a dilution did occur on consumption, then the maximum breath volatile intensity observed when consuming flavors dissolved in water, would not be expected to be very high, compared to the headspace concentration above the same solution. However, for many compounds (those with low air/water partition coefficients) the breath volatile concentration was found to be reasonably high compared to headspace: once the effects of absorption to the nasal epithelium, and dilution during transport through the upper airways had been taken into account (10). This suggested that dilution (sufficient enough to influence flavor delivery) did not occur earlier on in the release process either.

Flavor Delivery From Simple Solutions: Profile Intensity

It is also possible to examine the intensity of the flavor in the first peak (exhalation) after consumption and to compare this with static headspace data. The amounts of flavor delivered during consumption were different from those expected on the basis of headspace analysis (Figure 4). Headspace analysis showed very low concentrations of ethyl hexanoate in the gas phase above emulsions relative to those above water. In - vivo measurements showed that they were much more similar. It is possible to estimate the amount of dilution necessary to cause the change in flavor delivery. A 10 fold dilution of the bolus with saliva would be required for such a change in partition to occur (6). Such substantial dilution is unlikely, given the arguments outlined above.

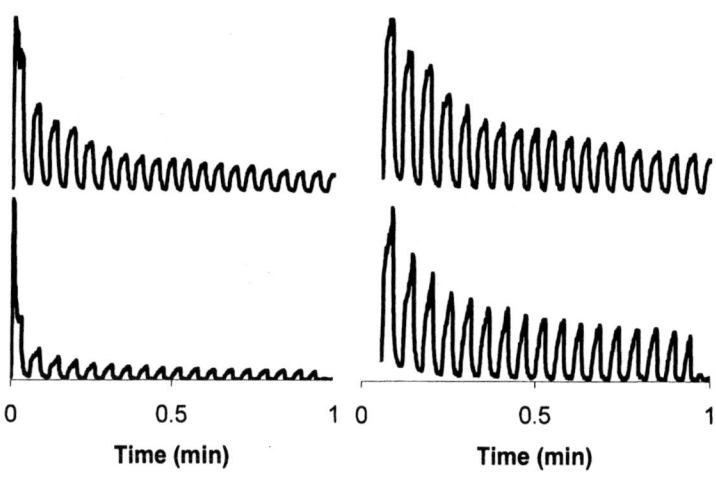

Figure 3. Breath by breath profiles from the consumption of solutions of ethyl octanoate in either a 2% lipid emulsion (upper) or water (lower). The profile on the left includes the first exhalation after swallowing (0 min). The profile on the right shows the second and subsequent exhalations magnified for comparison. Each profile shows the data from a single replicate, typical of those obtained during the analysis of a series of 5 replicates

Figure 4. Normalized release intensities of ethyl hexanoate in headspace and breath (in - vivo) when dissolved in either water or an emulsion (2% lipid) (Reproduced from reference 6. Copyright 2001 ACS)

The possibility remains that dilution may occur as a bolus is held in mouth during longer periods of mastication, compared to these experiments on solutions where the sample was swallowed immediately. However, although liquid systems were swallowed fairly rapidly, minimizing dilution, it is important to remember that dryer solid foodstuffs absorb saliva, reducing lipid saliva interactions. The primary effect of saliva in this instance, is to lubricate the bolus prior to swallowing. Indeed, during the eating process there is no real need to dilute a solid bolus into a dilute slurry, it is more important to form it into a softer cohesive mass for ingestion. If dilution by saliva does not appear to substantially affect flavor delivery, in either the short term, or the long term, then some other factor must account for the changes in flavor delivery.

The Effect Of The Partition Coefficient On Mass Transfer

The partition coefficient has been found to affect mass transfer both in dynamic headspace systems (7) and during consumption (10). In both instances the compounds with the lowest partition coefficients exhibited the best volatile transfer, relative to the amounts expected on the basis of static headspace studies. They had the greatest headspace stability, and consequently resistance to gas phase dilution, achieving close to static headspace breath concentrations upon consumption (when measured from the mouth, immediately after swallowing, to avoid loss due to nasal absorption).

The key factor in the dynamic headspace system model by Marin and co-workers (7) was the mass transfer coefficient. This varied between compounds according to eq 1; where k is the overall mass transfer coefficient, k_g and k_l are the mass transfer coefficients for the gas and liquid phases respectively and K_{aw}

is the air/water partition coefficient. It was only necessary to vary the value of K_{aw} in line with that of each compound, for the model to fit the data, k_g and k_l remained constant.

$$\frac{1}{k} = \frac{1}{k_g} + \frac{K_{aw}}{k_l} \qquad (1)$$

When solutions of flavor molecules in water were consumed a clear relationship between the air/water partition coefficient and in – vivo delivery was observed (Figure 5). Clearly compounds with lower partition coefficients were delivering concentrations closer to their static headspace than those with higher partition coefficients. This can explain the data presented in Figure 4. Ethyl hexanoate in water has a high partition coefficient and consequently delivers volatile inefficiently in - vivo, whereas ethyl hexanoate dissolved in an emulsion has a much lower partition coefficient and consequently delivers flavor more efficiently. Hence the two are more similar under dynamic conditions than at static equilibrium, although the absolute amounts released from the aqueous system will always be greater.

The change in mass transfer also explains the greater persistence of volatiles. The systems with lower partition coefficients have good mass transfer and more stable delivery relative to their maximum. In contrast, the high partition coefficient systems have poor mass transfer and as a consequence are less stable relative to their maximum, as seen for compounds with different K_{aw}'s dissolved in water (7).

Equally, it is also apparent that if dilution did occur and increased the partition coefficient, then delivery would become less efficient (Figure 5), negating part, if not all of the presumed benefit of increasing the partition coefficient in the first place. Consequently increasing the partition coefficient may not be the best mechanism for increasing flavor delivery under dynamic conditions.

Conclusions

Lipid has the potential for the retention of flavor molecules in foodstuffs. This can act as a reservoir, releasing flavor more stably at lower concentrations over the eating time course. This might be due to either the mass transfer characteristics of the food system (dependent on the partition coefficient itself), or to changes induced by dilution with saliva during consumption.

There was clear evidence that mass transfer had a substantial effect on flavor delivery in – vivo, and mass transfer was closely related to sample/air partitioning behavior. There was little evidence of a significant effect of sample

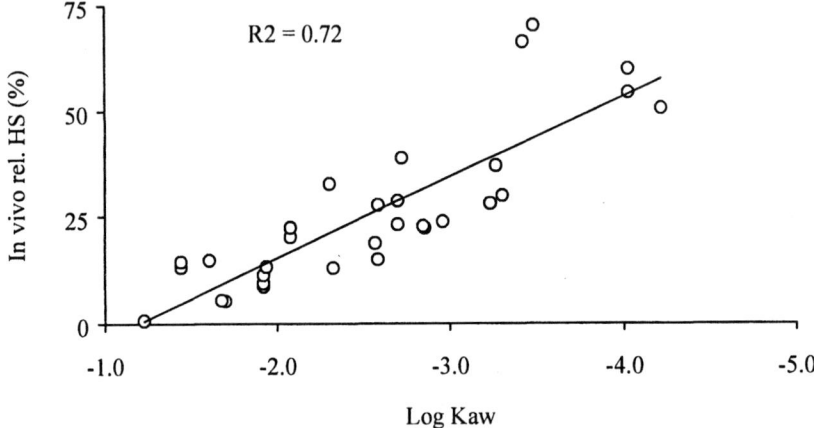

Figure 5. Relationship between the air water partition coefficient and in - vivo release (breath exhaled from the mouth: mouthspace) relative to static headspace measurements (Reproduced from reference 10. Copyright 2002 ACS)

dilution with saliva affecting flavor delivery. Mass transfer, driven by changes in the partition coefficient therefore appears to be one of the major factors affecting flavor delivery in lipid containing systems.

Acknowledgement

The authors are grateful to Firmenich SA Geneva for their financial support.

References

1. De Roos, K. B. *Food Technol.* **1997**, *51*, 60-62.
2. Brauss, M. S.; Linforth, R. S. T.; Cayeux, I.; Harvey, B.; Taylor, A. J. *J. Agric. Food Chem.* **1999**, *47*, 2055-2059.
3. McNulty, P. B.; Karel, M. *J Food Tech.* **1973**, *8*, 309-318.
4. Buttery, R. G.; Guadagni, D. G.; Ling, L. C. *J. Agric. Food Chem.* **1973**, *21*, 198-201.
5. Harrison, M.; Hills, B. P.; Bakker, J.; Clothier, T. *J Food Sci* **1997**, *62*, 653.
6. Doyen, K.; Carey, M.; Linforth, R. S. T.; Marin, M.; Taylor, A. J. *J. Agric. Food Chem.* **2001**, *49*, 804-810.
7. Marin, M.; Baek, I.; Taylor, A. J. *J. Agric. Food Chem.* **1999**, *47*, 4750-4755.
8. Medinsky, M. A.; Kimbell, J. S.; Morris, J. B.; Gerde, P.; Overton, J. H.. *Fundam. Appl. Toxicol.* **1993**, *20*, 265-272.
9. Buettner, A.; Schieberle, P. *Food Sci. Technol.* **2000**, *8*, 553-559.
10. Linforth, R.; Martin, F.; Carey, M.; Davidson, J.; Taylor, A. J. *J. Agric. Food Chem.* **2002**, *50*, 1111-1117.

Chapter 14

Fat Reduction in Foods: Microstructure Control of Oral Texture, Taste, and Aroma in Reduced Oil Systems

G. J. van den Oever

Unilever Foods Research Centre, Olivier van Noortlaan 120, 3133 AT Vlaardingen, The Netherlands

> Fat related sensory perception and consumer liking of food products often do not demonstrate a one-to-one relationship with fat level. This means that fat is involved in physical and/or perceptual mechanisms which can also be influenced by parameters independent of fat level. Control of these parameters is required in order to compensate for undesirable effects due to fat reduction. In order to realize the latter in a consumer perceivable way, oral physical properties have been identified, modeled, measured and validated which dominate sensory perception and preference of fat and fat replacement systems. Elements of sensory response obtained with mayonnaise (tasted as such) were found to be related to oral breakup efficiency and viscosity effects, and with spreads on bread being related to melting, water release, oil mass transport and lubrication. Models and mechanistic understanding allow prediction of sensory and/or consumer response from physical product characteristics as well as identification of superior fat replacers realized by microstructure control. Variations in fat level and microstructure in many systems also influence both texture and taste/aroma in a combined, simultaneous and inseparable way.

© 2006 American Chemical Society

Introduction

Sensory effects of reduced fat level play a role in many familiar food products, such as spreads [low (40 %) & very low fat (25 %)], cream alternatives, sauces/dressings, ice creams, and dairy spreads, among others. To control the consumer relevant part of these effects, a business model (given in Figure 1) has been proposed. In this model, physical and chemical product properties (as quantified at level 1) are related to consumer choice (as quantified at level 5) by identification and quantification of relevant relationships and effects at intermediate levels. We developed a method to realize the latter: integrated sensory (or consumer) response modeling (ISRM). In this chapter, this method and its results will be described and illustrated using examples from studies on fat reduction in foods focussing on mayonnaise and spreads on bread.

Explanation of the different levels in the business model
- Level 1 properties are the quantified physical and chemical product properties, just before oral processing. They are a microstructural function of ingredients (including fat level), processing and storage conditions.
- At level 2, the mechanisms of the relationships between these physical and chemical properties at level 1 and sensory properties at level 3 are studied: the oral physics level. Here, properties like oral deformability, particle size, flavor release and chewing parameters are quantified.
- At level 3 sensory properties are quantified, generally as the average result of a trained sensory panel.
- At level 4 product liking in a fixed context is quantified, either for the average of all consumers, or for the average per segment of consumers or even per individual consumer.
- At level 5 repeated product choice is measured, the most relevant business parameter.

Microstructure control starts with control of level 1 properties and will, under the right conditions, also control properties at higher levels in the business model. Dependent on critical conditions, control may extend to level 2, level 3 or, ideally even to level 5. Insight in these critical conditions is derived from the results of the ISRM-method as it delivers the complete set of physical/chemical drivers of a sensory attribute or liking in a mathematical equation and also identifies less relevant effects. Additional insight is derived, at the higher levels in the business model, from integration of the effects quantified at the lower levels with other consumer influencing parameters like context, expectations, costs and marketing. Ultimately, at level 5, the relative importance (dominance) of all elements of a given market proposition will co-define the scope of microstructural control.

Physiological variabilit.

It is clear that oral physical properties (level 2) will be a function of pre-oral product properties (level 1). It has to be realized, however, that also the settings of the oral processor will influence the level 2 properties: e.g. chewing and saliva production rate, number of chews, size and distribution of oral deformation forces and amylase enzyme activities. As large individual differences exist with respect to these oral settings, it can be expected that even if the food product served to a group of individuals has been very well defined in terms of its physical and chemical properties, the physical reality developed during consumption will demonstrate a considerable variability among individuals. This physical variability derived from one single food product may contribute to the sensory variability practically always encountered during sensory tests. Therefore, a statistical approach is required to derive scientifically relevant insights from sensory results in relationship to physical/chemical product properties and to identify and validate options for microstructure control.

Multi-modal interactions

Especially at the level of oral physics, food microstructures play a crucial role in controlling the oral delivery of the physical and chemical entities (stimuli) which will trigger the oral receptors such that texture and taste/aroma become a perceived reality in the brain of the consuming individual. In many studies which relate physical to sensory properties, multi-modal interactions have been encountered like effects of texture or color on taste/aroma perception. Part of these interactions will be a direct consequence of the oral physics and can be controlled by microstructural effects operating at level 2 in the business model. Another part of the interactions (e.g. effects of color on texture or aroma perception) has a basis of psychological association effective at the sensory level 3 and as a result of this in potential also at higher levels. Also some of these have been taken into account in our studies identifying the scope of microstructural control.

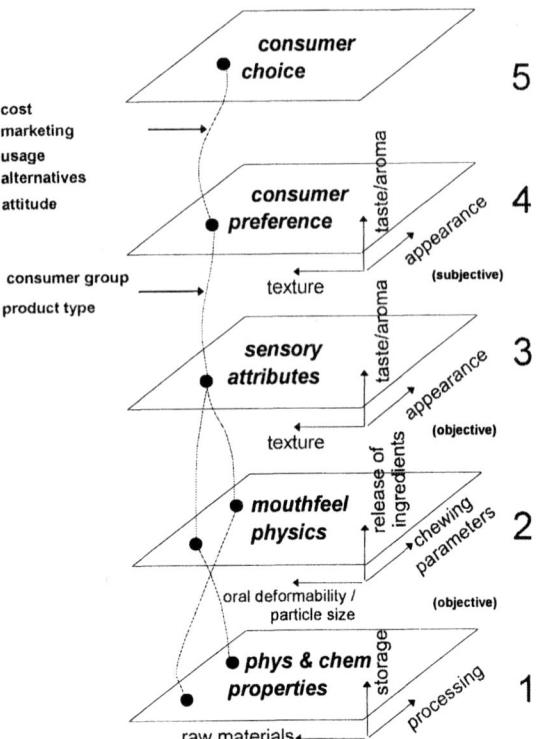

Figure 1. Business model relating effects of raw materials, processing and storage to 5 levels of food properties including consumer choice.

Integrated Sensory (or Consumer) Response Modelling (ISRM)

This methodology, introduced in reference 1, has been used to identify physical properties dominating consumer relevant sensory effects of fat reduction. Consumption test conditions either excluded chewing (e.g. studies on spreads, mayonnaise, dressings, sauces and soft cheese) or included chewing (f.i. studies on meat, cereals, vegetables, hard cheese, croissants, fish-fingers and spreads on bread or toast). In this chapter examples will be given of an ISRM-study without chewing (on mayonnaise) and an ISRM-study including chewing (of spreads on bread).

Chewing allows expression of physical effects of fat in the oral processes relevant to perception which are additional to the physical effects expressed without chewing. These effects may occur both at a product and a consumer level. At a product level, as an example could be mentioned fracture mechanical and lubrication effects on pieces of bread or pasta. These effects could be modified by the fat content in a spread used on top of the bread or a sauce used in combination with the pasta. They are additional to, for instance, effects related to oral emulsification which occur also without chewing. At a consumer level, individual chewing characteristics cause variations in physical effects, like shear rate and deformation force. These are per definition only operational and relevant during chewing and are additional to consumer related physical effects of salivary flow occurring also without chewing. Careful consideration of the physical effects operating during practical every day use of our foods by consumers is required to select the physical methods needed in ISRM-studies.

The ISRM-method applies the following general steps:
1. Identify sensory effects of fat replacement which have consumer relevance.
2. Carry out a qualitative analysis of the food physics relevant to the oral processes during eating.
3. Identify/define physical food properties which can be used to characterize these food physics.
4. Develop methodology to quantify these physical properties applying instrumental methods and/or models of principle.
5. Perform physical, sensory and (if possible) preference measurements on a wide and relevant range of products resulting in a homogeneous distribution of the products over the physical and sensory space.
6. Link measured sensory and/or preference properties to measured physical properties using statistical quantitative models. Derive from this analysis those physical properties which appear to be most dominant, operationally relevant and predictive for the sensory or preference properties studied.
7. Finetune and validate the models found using independent measures.

ISRM1: Identification of sensory effects of fat replacement which have consumer relevance

Early evaluation of fat reduction in foods caused negative consumer response related to sensory attributes like thick, sticky and creamy (for mayonnaise) and waxy, sticky, greasy, slimy, mouthcoating and filmlayer (for spreads on bread). Scored "oral lubrication" effectiveness of spread on bread as measured in 1987 decreased in the order of butter (80 % fat), jam (0 % fat), tub margarine (80), mayonnaise (80), monchou, sandwich spread (80), boursin, liver-paste, halvarine, wrapper margarine and peanut butter (80). This sequence indicates that orally perceivable effects which relate to lubrication of bread crumbs by spreads are not only a function of fat level, but can be influenced by other product properties as well. One of our challenges has been to identify these effects.

If we would like to link one of the aforementioned texture effects of fat replacement to level 4 in the business model (consumer liking), an awareness of covariant non-texture effects is very relevant for a correct interpretation of our results. Figure 1 (ex ref. 3) illustrates that by increasing the viscosity of an acid solution by addition of guar gum (which clearly results in a perceivable increased thickness) also the acid intensity perception is reduced. Any consumer response on these systems in terms of preference can only be interpreted as a reaction on a combined taste and texture effect and not on one of each separately.

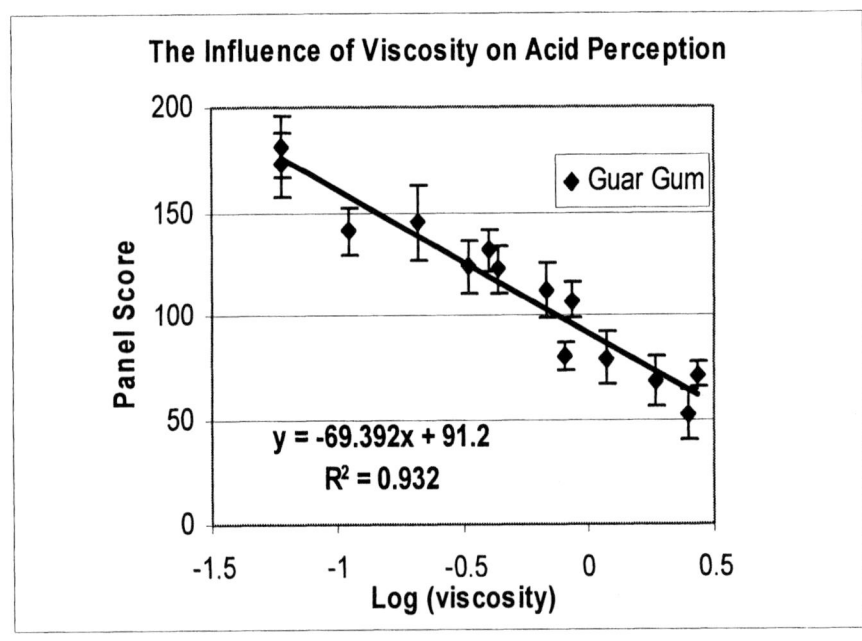

Figure 1. Texture and taste effects of guar thickening

Another example of this phenomenon was obtained in a study performed in cooperation with TNO-Foods in Zeist, The Netherlands. After identification of physical parameters driving consumer liking of spreads on bread under a given set of test conditions, a model was derived using linear regression (PLS) relating the combined physical effect in the liking model to sensory attributes obtained with a quantitative descriptive sensory panel. This was done to obtain information about the sensory effects involved in the physical effects on liking. The combined physical effect in the liking model is called "physics boosting liking". It consists of the mathematical equation in which all physical parameters (physical drivers of liking) are represented, together with their appropriate coefficient, which were validated to dominate (together) consumer liking. The coefficients for each of the sensory attributes in the model relating this "physics boosting liking" to sensory attributes are given in Table I. A high positive value means a high positive correlation to the physics boosting consumer liking; a high negative value means a high negative correlation to the physics boosting liking. It is clear that not only texture related effects like roughness, stickiness and creamy mouthfeel correlate to this overall effect of the physical drivers of liking but also taste and aroma attributes like salty taste and margarine odor.

Summarizing, it can be expected that microstructural control in fully formulated products will most often involve parallel, inseparable effects both on texture and on aroma and taste perception.

Table I. Coefficients of sensory attributes underlying the effect quantified by the physical parameters in best spreads on bread liking model (from PLS on standardized values)

Attribute	Coefficient
ROUGH	-0.057
CHEW FORCE	-0.056
CHEW TIME	-0.051
STICKINESS	-0.048
STICKY AFTERFEEL	-0.048
TONGUE MOVEMENTS	-0.039
TOUGHNESS	-0.035
MARGARINE ODOUR	0.042
SQUEEZE	0.047
AIRY	0.049
OILY AFTERFEEL	0.051
CREAMY MOUTHEEL	0.054
SALIVA	0.054
FULL TASTE	0.055
SALTY TASTE	0.059

ISRM2. Qualitative analysis of the physics relevant to the oral processes during consumption of the spread/bread system

The results of a qualitative analysis of 17 macro- and micro-scale physical processes which may have relevance to the sensory perception of fat level effects in the spread/bread system are given below:

1. *Pretreatment of bread:* Toasting will affect the consistency, hydrophobicity and temperature of the substrate brought into contact with the spread. A high temperature will boost pre-oral spread melting (process 3: see below). In some of our research on spread fat level effects, effects of pretreatment conditions have been studied as one of the consumer relevant variables. In the majority of our studies pretreatment conditions, including storage conditions of the substrate, have been standardized.
2. *Covering bread with spread:* Spread hardness will influence spread dosage and layer thickness homogeneity. Spreading forces will be adapted to perceived resistance thereby influencing the expectations anticipating the following oral consumption. In most of our studies the expectation effect was ruled out by having the spread/bread system prepared by the experimentalist. Also the effect of spread hardness on spreading efficiency was normalized by providing constant dosage and layer thickness.
3. *Pre-oral melting after spreading*, promoting process 4 below.
4. *Pre-oral spread penetration into the bread pores* induced by spreading forces and affected by melting. Less hard, better melting products will penetrate the pores to a larger extent, thereby promoting process 5.
5. *Oral melting of the spread*, promoting process 6: Independent of bread pretreatment all spreads will reach a temperature between 30 and 35°C in the mouth.
6. *Water suction from (watercontinuous) spread*: Pre-oral and oral wetting of the bread/toast may occur by spread water sucked out of a water continuous spread by capillary action. This process will soften the crumbs, influence crumb size reduction (process 8) and decrease the need for saliva wetting of the crumbs (process 12), thereby increasing the amount of saliva available for crumb friction reduction (process 14). The effect will also influence processes 11-16 to the extent that the spread residue remaining after water release will have a higher viscosity. The sensory effect of this function has been studied by varying the spread/bread contact time before oral processing. Increasing toast hydrophobicity by using extended toasting times will counteract wetting and change the relevance of the spread water release process.
7. *Swelling of (watercontinuous) spread*: Instead of releasing water to the toast, a water-continuous spread might take in saliva fluid due to an osmotic

difference. This could reduce the oral viscosity of the spread and thereby fasten all processes involving oral spread transport (processes 10 and 15).

8. *Oral particle size reduction:* This initial goal of the chewing process is influenced by bread fracture properties (more or less spread modified) and biophysical differences among consumers. The resulting increase in crumb surface area will boost process 10 (surface coverage with spread), process 6 (water suction from spread) and process 12 (saliva wetting of crumbs). It will increase the need for process 14 (crumb friction reduction).

9. *Bread glueing:* The crumb surface area available for saliva wetting will be decreased by adhesive properties of initially wetted toast. Water which may be released from the spread, will increase the stickiness of the crumbs.

10. *Surface coverage*: enlargement of the spread/bread interface area in the mouth is induced by mixing forces and influenced by crumb size and spread flow resistance as function of temperature (process 3), shear rate and reactions with saliva. A lower spread viscosity will lead to a larger interfacial area, thus influencing the processes 11-15. This effect will be influenced by biophysical variations among consumers causing differences in shear rate profiles in the oral cavity.

11. *Spread glueing*: The crumb surface area available for saliva wetting will be decreased by adhesive properties of spread present between 2 crumb surfaces. The latter property will be influenced by spread viscosity and stability. Also a viscosity increase due to water suction from the spread into the toast will boost the spread glueing. Consumer complaints in the area of "sticky bolus / doughball" perception (chapter 1) are expected to be related to the processes 9 (bread glue), 11 (spread glue) and 14 (crumb friction reduction).

12. *Saliva wetting of the bread/toast:* This process operates both through fast capillary suction and slow diffusion. It will be influenced by biophysical differences between consumers wrt. their saliva productivity. This process counteracts process 14 (crumb friction reduction) and will be influenced by process 10 (surface coverage with spread). Capillary suction is counteracted either by oil making the pore surface area hydrophobic or by viscous structured water filling the pores as cement.

13. *Amylase action:* Bread/toast starch breakdown will decrease crumb hardness cq. size, thicken the saliva and possibly increase the bread glue (process 9). Bread surface coverage by spread (process 10) will influence the kinetics of amylase breakdown and thus the practical relevance of possible negative glue effects.

14. *Crumb lubrication:* This final effect will influence the effective bolus deformability. It will increase with decreasing crumb volume fraction (less

saliva and less spread inside the crumbs) and decreasing friction coefficient of the phase between the crumbs (saliva best; oil next; saliva thickened with bread starch broken down by amylase intermediate; shear thinned and/or saliva-dispersed structured water from spread intermediate; non-thinned / non saliva-dispersed structured water from spread worst).

15. *Loose spread transport:* This process will gain importance to the extent that the spread loses its contact with the bread (the opposites of process 4: spread penetration into bread pores, and process 10: bread surface coverage with spread). Loose spread will be more likely to be swallowed in an early stage of the chewing process, thus losing its possibility to contribute to process 14 (crumb friction reduction). However, if this lose spread is not swallowed early it may have a very positive effect by "outer lubrication" of a coherent bolus which consists of internally less lubricated crumbs. The lose spread effect could be rather sensitive to biophysical differences between consumers and be one of the factors explaining variations in consumer response on one single product. The extent of lose spread transport will be an important parameter in process 16 (mouthcoating). Loose spread transport will increase with increasing the spread/bread ratio.

16. *Mouthcoating:* This effect will both increase with increasing extent of loose spread transport (process 15) and decrease to the extent that oral shear forces are able to scrape off layers adsorbed on the oral tissues. In general, scraping off will increase with decreasing oral spread viscosity.

17. *External Bolus Lubrication:* Even if a bolus suffers from considerable internal friction and/or consists of crumbs which are glued to each other, the outside of such a low deformable bolus could be well lubricated by a suitable spread/saliva mixture. In an extreme illustration of this functionality, part of the spread would release water to hot toast and be thickened such that the crumb surfaces affected would be glued and poorly lubricated, thereby forming a low deformable bolus, whilst another part of the spread would be separated into the oral cavity, be mixed and thinned with saliva and then act as an external lubricant. Also this effect would be very sensitive to biophysical variations. Expression of the effect would increase with increasing spread/bread-ratio.

Based on and derived from the upper analysis many physical properties have been measured on pretreated bread, spreads as such and the spread/bread combination. The results have been used to find by ISRM the sensory dominant processes, properties and fat effects relevant to the food system under study.

ISRM3-4. Measurable physical properties relevant to oral processes involved in mayonnaise consumption

Based on an analysis of pure mayonnaise consumption comparable to the one given in the former chapter, the following measurements were carried out on a range of mayonnaise systems:

- Lubrication traction coefficient from tribology (see reference 2)
- Deformability from
    - Stevens values at 5 and 20 °C
    - Viscoelastic properties measured in oscillatory strain sweep at 1 Hz:
    - Storage (G') and Loss (G") Modulus
    - Critical (i.e. G' = G") Strain and Stress
    - Tg Delta (G"/G') at strain 0.001
- D_{32} oil droplet size
- Fat Level
- Specific Volume
- Effect on flow resistance of dilution with artificial saliva
- Effect on flow resistance of dilution with artificial saliva + amylase
- Oil deposition on a mucin film (reference 2)
- Flavor release using MS-breath (reference 3)

ISRM5. Experimental characterization of wide ranges within a food product category

Figure 2 displays a selection of physical properties based on the earlier analysis as well as consumer preference data for a range of mayonnaises. Figure 3 gives an analogue set of data for spreads on bread in relationship to sensory fattiness. Both figures suggest that consumer segment preference for mayonnaise as well as sensory fattiness of the spread/bread system are influenced by additional factors than fat level only. It is also clear that many physical properties exist which have a relationship with oral processes but are not a simple function of fat level. More experimental data obtained with the spread/bread system are given in reference 1.

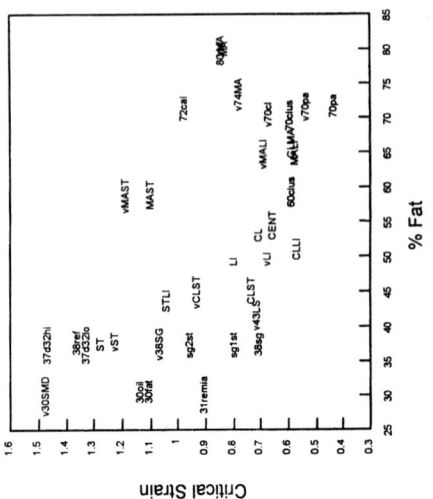
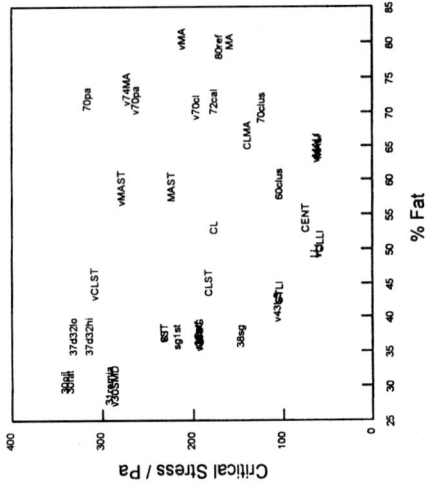

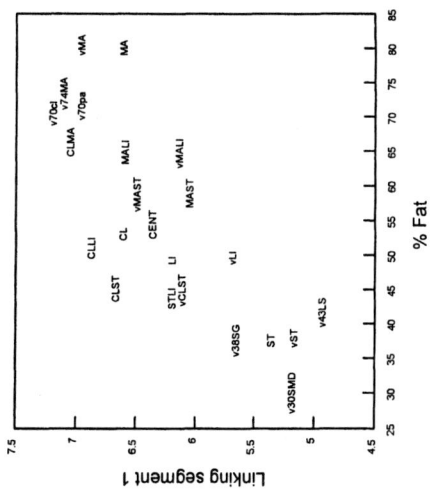
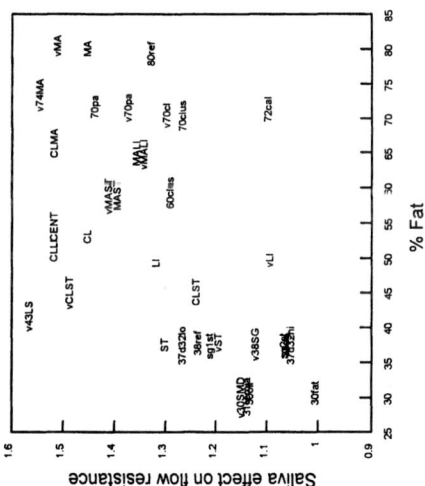

Figure 2. Selection of physical and consumer preference data obtained for a range of mayonnaises, as function of fat level

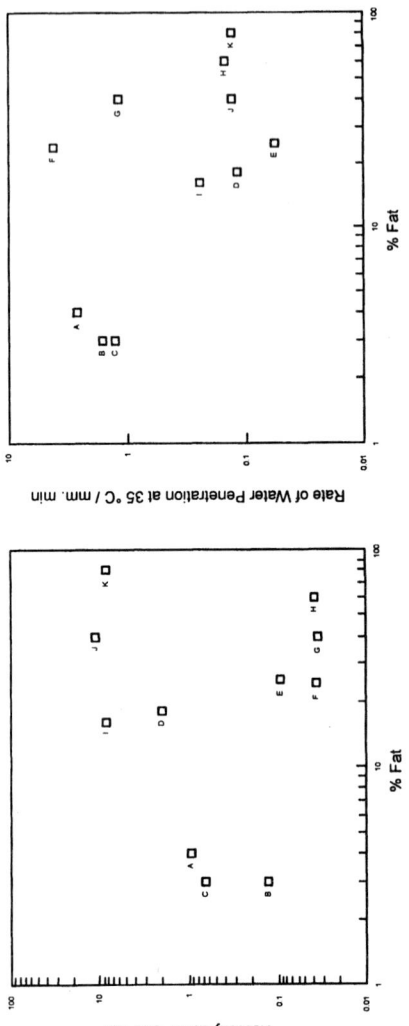

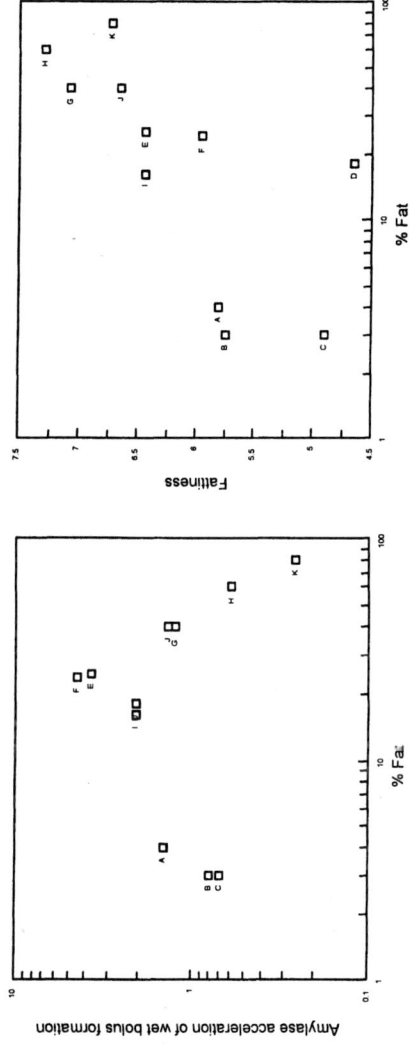

Figure 3. Selection of physical and sensory fattiness data obtained for a range of spread/bread systems, as function of fat level

ISRM6. Quantitative models of sensory or consumer preference data

Linear regression procedures were performed linking physical to sensory and preference data, similar to those given in Figure 2 and Figure 3. By doing so, those physical properties were selected which explain the measured sensory and consumer preference variance such that the highest levels of statistical significance per explanatory variable were combined with maximum predictability of measured samples kept out of the analysis (cross-validation). As an additional selection criterion different physical properties within one model should have low Pearson correlation coefficients (mutual independency). The dominant, operationally relevant and predictive physical properties thus selected could be considered as the drivers of the sensory or preference properties studied (see the example in Figure 4).

ISRM7. Validation contributing to mechanistic understanding and microstructure control

As illustrated by the scheme in Figure 5, validity of any of the ISR-models found, can be derived from their predictability of the sensory scores or consumer preference data obtained with new samples and/or consumers. A high predictability confirms that indeed relevant and dominant physical parameters have been identified: the real drivers. Additionally, the way in which the physical parameters contribute to the model (1) needs to be considered in combination with (2) the measured fat effects on each of these physical parameters (as f.i. in Figure 2 and Figure 3) and with (3) additional microstructure characterization of the systems which influence these physical properties independent of fat level. This 3 factor integration generally allows clear hypotheses to be tested for the mechanistic basis of the observed fat effects on sensory or preference parameters. It generates information about the relationship between microstructure and sensory dominant physical properties and, ultimately, reveals the possibilities for the microstructure control we aim for.

The following mechanistic understanding has been obtained for mouthfeel differences obtained with 3-60 % fat spreads and 16-80 % dressings on hot white toast. These mouthfeel effects can be explained by spread differences working through 4 independent mechanisms: melting, water release to toast, oil lubrication and oil mass transport (location / kinetics). Also, information on their relative importance has been obtained. This understanding has been based on ISR-models relating mouthfeel differences to differences in physical parameters

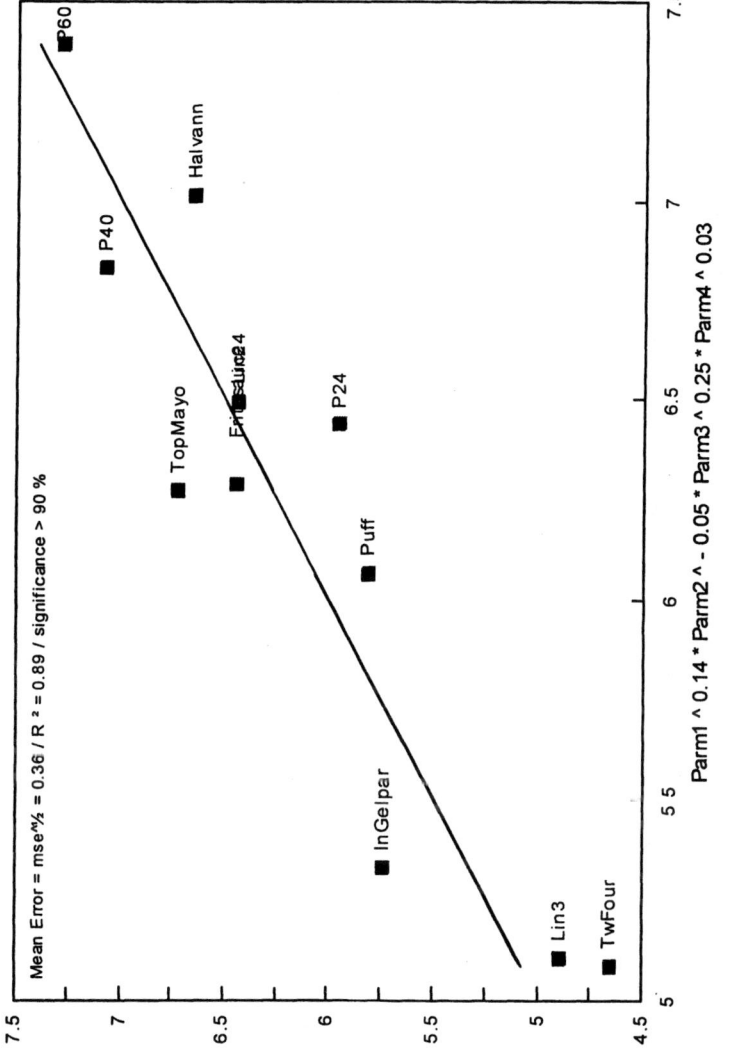

Figure 4. Quantitative model identifying drivers of sensory fattiness of the spread/bread system

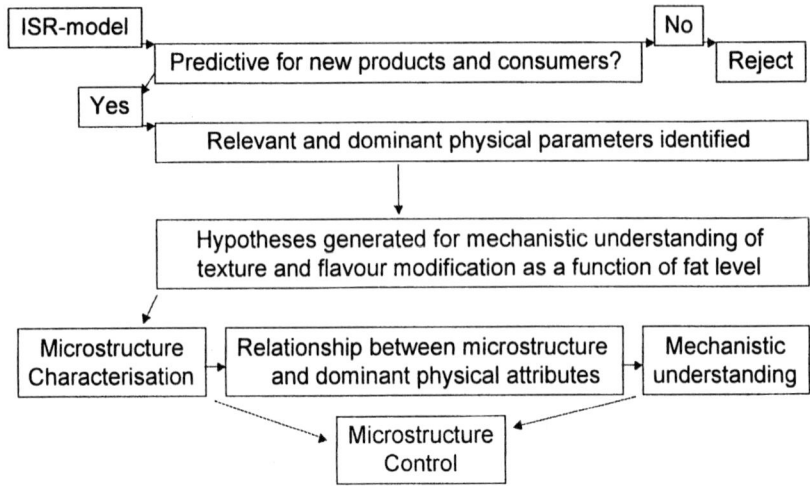

Figure 5. Scheme of validation of ISR-models contributing to microstructure control

measured both on the pure spread and the spread/bread-system. During the ISRM-process also spread effects were identified showing no relevance, or considerably less dominance, for the sensory window studied. Thus, an improved focus in designing novel ways of microstructure control has been obtained as these will predominantly work through (combinations of) the driving mechanisms identified. In pure mayonnaise consumption these driving mechanisms appeared to relate to oral break-up efficiency and viscosity effects.

Conclusions

A method has been developed to study sensory perception and consumer acceptance of low fat products in the context of a 5-level business model for fat effects. This method, integrated sensory (or consumer) response modeling (ISRM), provides insight in the conditions under which microstructure control of consumer relevant sensory perception is feasible. It identifies physical/chemical drivers of sensory perception or consumer liking and quantifies the effect of microstructure control of these driving physical/chemical properties, relative to other practical effects on sensory perception or liking. The method also provides hypotheses or supportive evidence for mechanistic understanding of the oral processes dominating the sensory response under study.

Physical properties have been identified, modeled, measured and validated which dominate sensory perception and liking of fat and fat replacement systems in mayonnaise and spreads on bread. Elements of sensory response obtained with mayonnaise (tasted as such) were found to be related to oral breakup efficiency and viscosity effects, and with spreads on bread to be related to melting, water release, oil mass transport and lubrication.

Models and mechanistic understanding allow prediction of sensory and/or consumer response from physical product characteristics as well as identification of superior fat replacers realized by microstructure control. Variations in fat level and microstructure in many systems influence both texture and taste/aroma in a combined, simultaneous and inseparable way.

References

1. van der Oever, G.J. *Cereal Chem.* **2003**, *80*, 409-418
2. Malone, M.E. Appelqvist, I.A.M. and Norton, I.T. *Food Hydrocoll.* **2003**, *17*, 763-773
3. Malone, M.E. Appelqvist, I.A.M. and Norton, I.T. *Food Hydrocoll.* **2003**, *17*, 775-784

Chapter 15

Effect of Composition of Triacylglycerols on Aroma Volatility: Application to Commercial Fats

Natacha Roudnitzky[1], Gaëlle Rondaut[*,2], and Elisabeth Guichard[1]

[1]Institut National de la Recherche Agronomique, Unité Mixte de Recherches sur les Arômes, 17 rue Sully, 21065 Dijon, France
[2]Ingénierie Moléculaire et Sensorielle des Aliments et des Produits de

Santé, ENSBANA, 1 Esplanade Erasme, 21000 Dijon, France

The effect of pure triacylglycerols of different compositions (chain-length and unsaturation level of fatty acids) on the release of flavor compounds was determined at different temperatures above their melting points. A linear relationship was established between the volatility of the flavor compounds, on one hand, and the triacylglycerol fatty acid chain length and degree of unsaturation, on the other hand. In order to have a general relation, which applies also to complex lipids, triacylglycerol fatty acid chain length and degree of unsaturation were respectively replaced by weighted averages of complex lipid chain-length and degree of unsaturation. Applications are presented for complex lipids and emulsions.

Since most flavor compounds are highly soluble in fat, changing the fat content of a food product may cause a change in the flavor release from the matrix, and consequently, a noticeable change in its perception. The influence of fat content has already been extensively studied, whereas no such fundamental work has been done on the effect of fat nature on flavor release (*1*).

Welsh and Williams (*2*) found a limited effect of oil type on oil / water partition coefficients of volatile compounds from different chemical classes. However, Vedovati (*3*) showed a relation between oil hydrophobicity and allyl isothiocyanate oil / water partition coefficients. Maier (*4*) showed that volatility of solvents depends on both triacylglycerols fatty acid chain length and the degree of unsaturation of fatty acid methyl esters. More recently, Guichard *et al.* (*5*) reported a negative correlation between the level of unsaturation in oils and the release of hydrophobic flavor compounds.

The aim of the present study is then to focus on chain length effect and unsaturation level effect on flavor release, with pure triacylglycerol and with complex lipids, when in liquid state.

Experimental

Model and complex lipids

The model lipids chosen were triacylglycerols: tributyrin (C 4:0), tricaprylin (C 8:0), trilaurin (C 12:0), trimyristin (C 14:0), tristearin (C 18:0), and triolein (C 18:1). They were supplied by Sigma-Aldrich (Saint Louis, MO), of analytical grade and 95 % pure. Tristearin was only of 60% purity.

Two sets of complex lipids were studied. The first one was composed of commercial fats with different melting points. They were one vegetable fat, a coconut oil from *Cocos nucifera* (Sigma-Aldrich), and three animal fats, three deodorized Anhydrous Milk Fats (Corman SA, Goé, Belgium) with melting points of 20, 32 and 41°C.

The second set was composed of complex lipids, all in liquid state at ambient temperature. Their choice was based on their differences in fatty acids composition. They were vegetable oils : four edible oils, olive oil, sunflower oil, rapeseed oil and oleic sunflower oil delivered by Lesieur SA (Neuilly-sur-Seine, France), and a non-edible oil, linseed oil (VWR International Inc, West Chester, PA). All lipids were stored at 4°C under nitrogen and in the dark.

Lipid characterization

DSC thermograms

Aliquots of 5 to 10 mg fat were put into aluminum sealed pans. Differential Scanning Calorimetry was performed on a DSC 7 (Perkin Elmer Inc, Wellesley, MA), calibrated with water and n-octane. Triacylglycerols were melted prior to the analysis and the DSC pans were heated from –120°C to 80°C, at 10°C.min^{-1}.

Complex lipids composition: fatty acids profile

Isopropyl esters extraction: Isopropyl esters were obtained by lipids transesterification (6).

When required lipids were melted beforehand, then 3 to 6 mg lipid were weighted out accurately in a centrifugation vial. Samples were dried under nitrogen (N-EVAP 111 ; Organomation Associates Inc, Berlin, MA, USA), and volumes of 0.5 mL hexane, 3.5 mL isopropanol, and 0.5 mL sulfuric acid were added. The samples were hermetically closed, shaken with a vortex and stored in a steam room at 100°C for 1 hour.

Volumes of 0.5 mL 5% NaCl solution, 1.5 mL hexane and 5 mL distilled water were added to the sample at room temperature. The samples were shaken with a vortex and centrifuged at 3000 rpm.min^{-1} during 3 min. Organic phase, which contained isopropyl esters, was removed, and put in vials. A volume of 2 mL hexane was one more time added for a second extraction of the organic phase. Both organic phases were put together, and then diluted with hexane in a 2 mL vial, to obtain an isopropyl esters estimated concentration between 0.5 and 1 µg.µL^{-1}.

Methyl esters extraction : Methyl esters were prepared as described elsewhere (7). Between 1 and 20 mg oil were weighted accurately in a centrifuge tube. Samples were dried under nitrogen (N-EVAP 111 ; Organomation Associates Inc), and 0.3 mL toluene and 0.7 mL 7 % BF3-methanol (Supelco Inc, Bellefonte, PA) were added. Samples were hermetically closed under nitrogen atmosphere, and stored in a steam room at 95°C for 2 hours.

Two milliliters of hexane and 5 mL saturated solution of NaHCO$_3$ were added to the sample at room temperature. Samples were shaken with a vortex and centrifuged at 3000 rpm.min^{-1} during 3 min. Organic phase, which contained methyl esters, was removed, and put in vials. Hexane was evaporated under nitrogen flow at 40°C during 15 min. The dried sample was again diluted in hexane to obtain a methyl esters at a concentration of 0.5-1 µg.µL^{-1}.

Gas chromatographic analysis

One microliter of the previous solution was injected into a fused silica capillary column DB-WAX (30 m x 0.32 mm id, film thickness 0.25 µm; Agilent Technologies, J&W Scientific brand, Palo Alto, CA). Injections were made in the splitless mode with a split time of 30 s into a Hewlett Packard 6890 gas chromatograph coupled to a Hewlett Packard 5973 series mass-selective detector (Agilent Technologies). Pressure was kept constant at 0.44 bar, to ensure a helium gas flow of 40 $cm.s^{-1}$ at 110°C. The temperature program for isopropyl esters included heating from 50°C to 190°C at 5°C.min^{-1}, a 15 min steady step at 190°C, followed by a heating from 190°C to 240°C at 20°C.min^{-1}, and 15 min at 240°C. For methyl esters, the temperature program included a heating stage from 50°C to 240°C at 5°C.min^{-1}, and then a steady stage at 240°C.

Emulsions

Oil in water emulsions were prepared as described by Espinoza-Diaz (8) and Charles et al. (9) experiments; they were made of 30 % oleic phase (olive, oleic sunflower, rapeseed, sunflower or linseed oil) and 70 % aqueous phase, a solution of 0.71 % β-lactoglobuline (Davisco Foods International Inc, Eden Prairie, MN) adjusted at pH 3 with 1 M HCl.

Flavor compounds were ethyl hexanoate (Sigma-Aldrich) and allyl isothiocyanate (Sigma-Aldrich), they were of analytical grade and 95 % pure. They were added in the oleic phase in the non-emulsified systems at a concentration from 1.5 to 3 $\mu l.g^{-1}$. After one hour equilibrium, samples were emulsified with a homogenizer Polytron PT 3100 (Fisher Scientific International Inc, Hampton, NH) at 15 000 rpm.min^{-1} over 5 min.

The lipid particle size distribution of the emulsions was controlled by laser light granulometry, with a Mastersizer 2000 (Malvern Instruments Ltd, Malvern, Worcestershire, UK) equipped with a Hydro 2000G tank.

Static headspace analysis

Samples, lipids and emulsions, were distributed into 20 mL vials (5 g per vial) and when required, lipids were beforehand melted. Flavor compounds, the same as in emulsions (ethyl hexanoate and allyl isothiocyanate) were added at the concentration of 1.5 to 3 $\mu l.g^{-1}$ to the lipid matrices. Thermodynamic equilibrium was obtained after 3 hours, or 1 hour 30 min for, respectively, lipid matrices and emulsions. The analyses were carried out at different temperatures: 15C, 30C, 45C or 60°C.

Headspace samples of 1 mL were taken by a multipurpose sampler 2 (Gerstel GmbH & Co.KG, Mülheim an der Ruhr, Germany), and injected into a fused silica capillary column DB-WAX (30 m x 0.32 mm id, film thickness 0.25 µm; Agilent Technologies). Injections were made in the split mode with a split ratio of 4.5 : 1 onto a Hewlett Packard 5890 gas chromatograph coupled to a Hewlett Packard 5970 series mass-selective detector. Helium carrier gas had a velocity of 40 cm.s^{-1}, temperature was kept constant at 100°C or 115°C, respectively for ethyl hexanoate and allyl isothiocyanate.

Statistical analysis

Statistical analyses were realized on data of flavor compounds chromatographic surface area in the gaseous phase, using SAS software (SAS Institute Inc, Cary, NC). Analyses of variance were made with the procedure GLM, followed by a Student Newman Keuls test with a risk of 5 %; linear and non-linear regressions were respectively made with the procedure REG, and the procedure NLIN.

Results and Discussion

Effect of triacylglycerol composition on aroma volatility

Characterization of triacylglycerols melting temperatures

Differential scanning calorimetry allowed detemrination of the melting range of the lipids : determining both the onset and the end temperature of the melting (Figure 1).

The triacylglycerol chain-length increases from tributyrin (C 4:0) to tristearin (C 18:0), and all triacylglycerols have no unsaturation, except triolein (C 18:1) with one unsaturation.

At 15°C and at 30°C, only tributyrin, tricaprylin and triolein are in liquid state. At 45°C, trilaurin is also liquid and at 60°C all triacylglycerols are in liquid state.

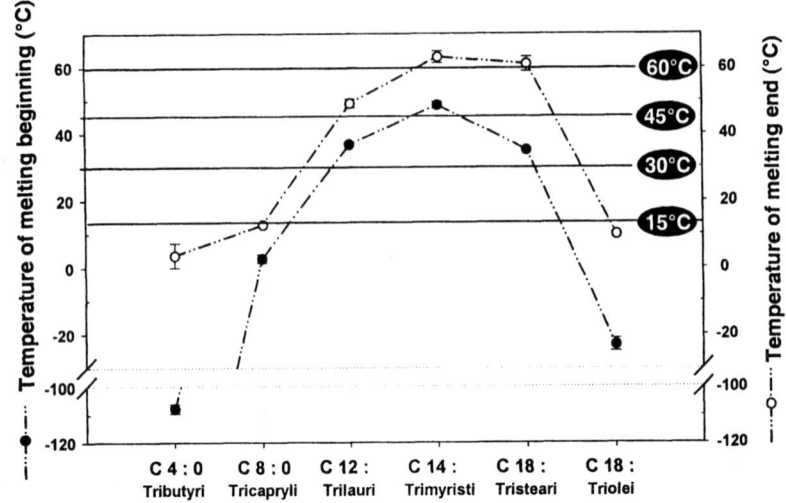

Figure 1. Representation of temperature range for the melting of the triacylglycerols studied

Effect of temperature on aroma volatility

Flavors release was studied when triacylglycerols are in liquid state. The chromatographic surface areas of allyl isothiocyanate, measured as a function of temperature, are plotted on figure 2.

For each triacylglycerol, allyl isothiocyanate chromatographic surface areas increase with the temperature, and the data can be fitted with exponential laws for tributyrin, tricaprylin and triolein. These fittings confirm the application of the Van't Hoff law to lipid systems, which have low air / oil partition coefficients. The same phenomenon is observed with ethyl hexanoate.

Effect of triacylglycerol fatty acid chain length on aroma volatility

The previously presented data can also be represented as a function of carbon number at different temperatures (Figure 3).

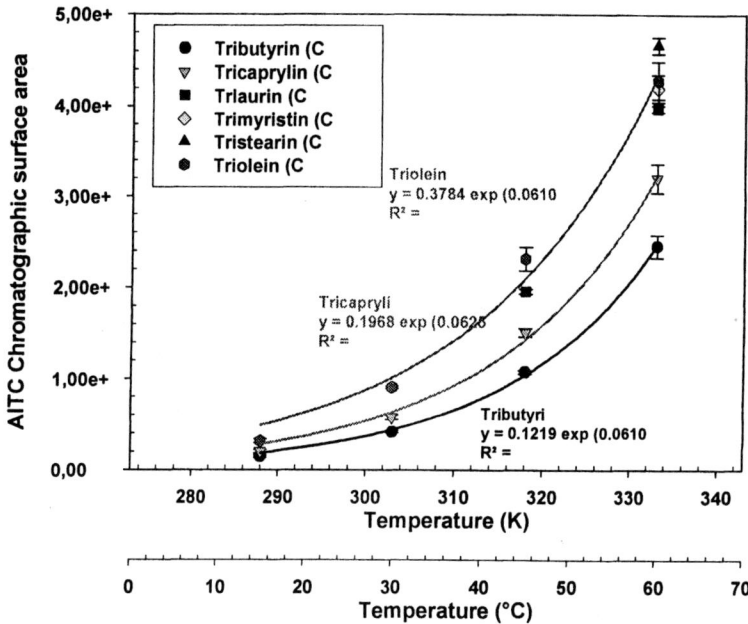

Figure 2. Allyl isothiocyanate chromatographic surface area as a function of temperature

For both flavor compounds, there are linear relations between chromatographic surface areas and carbon number of saturated triacylglycerols at 45°C and at 60°C.

It is noticeable that the values of the slopes are higher for allyl isothiocyanate than for ethyl hexanoate. However, due to the unsaturation, significant differences exist between chromatographic surface areas in triolein and the values predicted by the linear relationships for saturated triacylglycerols.

Mathematical linearization

Linear relations between triiacylglycerols carbon number and chromatographic surface area for different temperatures were determined, but these relations were not valid for unsaturated triacylglycerols. Moreover, Guichard *et al.* (5) found a linear relationship between calculated oil degree of unsaturation and the decrease in flavor release of allyl isothiocyanate and ethyl hexanoate.

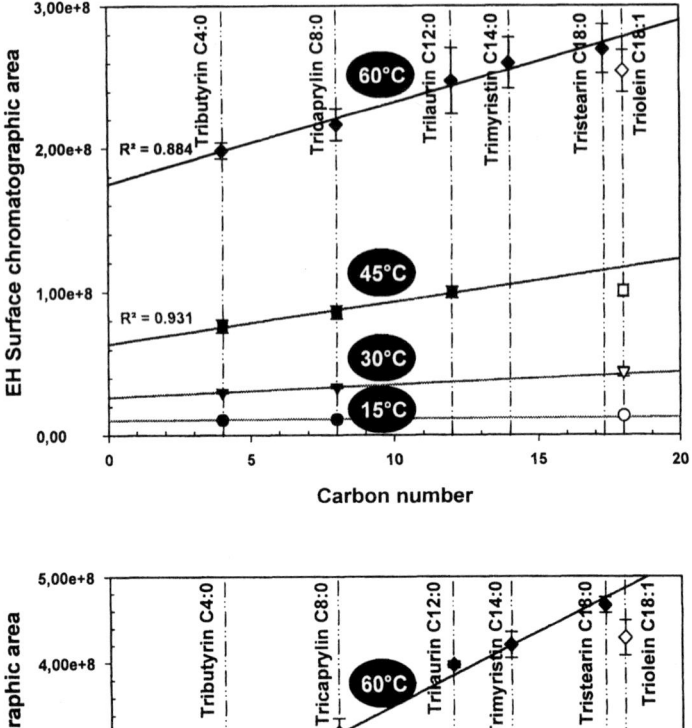
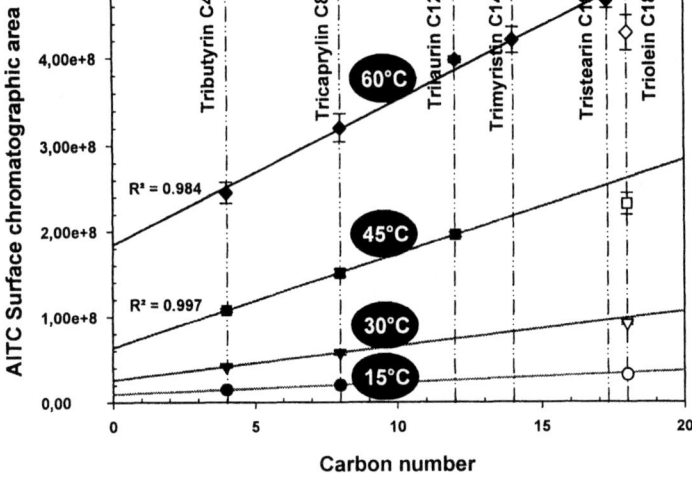

Figure 3. Ethyl hexanoate (top) and allyl isothiocyanate (bottom) chromatographic surface area as a function of saturated triacylglycerol chain length. Linear regression was only calculated for saturated triacylglycerol

We then suggested to fit experimental data with the following linear relationship, taking into account both variables :

$$CS = (a \times CN) + (b \times DU) + k \tag{1}$$

CS = Chromatographic Surface area
CN = Carbon Number chain length; DU = Degree of Unsaturation

Table I. Application of the model to flavor release in triglycerides.

Ethyl hexanoate					
Temperature	a	b	k	r^2	$Pr > |t|$
15°C	106 804	1 424 866	<u>10 081 995</u>	0.788	0.0009
30°C	<u>877 173</u>	1 747 221	<u>26 393 512</u>	0.967	<.0001
45°C	<u>2 838 099</u>	<u>-15 923 658</u>	<u>63 580 577</u>	0.942	<.0001
60°C	<u>5 747 183</u>	<u>-23 852 864</u>	<u>174 684 800</u>	0.877	<.0001
Allyl isothiocyanate					
Temperature	a	b	k	r^2	$Pr > |t|$
15°C	<u>1 377 008</u>	-2 108 004	<u>9 216 832</u>	0.981	<.0001
30°C	<u>4 000 398</u>	<u>-6 883 813</u>	<u>25 823 933</u>	0.998	<.0001
45°C	<u>10 986 636</u>	<u>-29 806 813</u>	<u>63 596 814</u>	0.993	<.0001
60°C	<u>16 792 426</u>	<u>-58 044 427</u>	<u>184 182 273</u>	0.083	<.0001

Underlined parameters are significant with a risk of 5 %;
$a = a' \exp(a'' \times T) ; b = b' \exp(b'' \times T) ; k = k' \exp(k'' \times T)$.

It is noticeable that, at 15°C for ethyl hexanoate, due to the small number of triacylglycerols in liquid state (tributyrin, tricaprylin and triolein), r^2 is relatively low and only the value of the intercept is significant.

At 30°C for ethyl hexanoate and 15°C for allyl isothiocyanate, only parameter b for degree of unsaturation is not significant. In the other cases, the model can be applied, and all parameters are highly significant.

Applications to complex lipids and emulsions

In order to apply the model to complex lipids, the triacylglycerol carbon number and degree of unsaturation were respectively replaced by weighted averages of complex lipids carbon number and degree of unsaturation. Experimental data were fitted with equation (1), and parameters a, b, and k were calculated.

First application : commercial fats with different melting points

Fats were beforehand characterized in order to calculate weighted average of carbon number and degree of unsaturation (Table II).

Table II. Commercial fats composition: fatty acids profile (GC area %)

Fatty acids	AMF-41°C	AMF-32°C	AMF-20°C	Coconut oil
C 4:0	4.14	5.04	5.55	-
C 6:0	2.44	3.45	3.61	-
C 8:0	1.56	1.93	2.03	8.51
C 10:0	3.67	3.89	3.93	6.57
C 12:0	4.82	4.75	4.36	**48.14**
C 14:0	**14.67**	**12.95**	**12.27**	**16.35**
C 16:0	**37.92**	**33.78**	**28.21**	7.77
C 16:1	1.19	1.45	1.66	-
C 18:0	**13.74**	**10.41**	**11.06**	2.31
C 18:1	**15.92**	**20.31**	**26.36**	5.96
C 18:2	-	-	-	2.45
C 18:3	-	1.58	-	1.15

Major saturated fatty acids in anhydrous milk fats are myristic (C 14:0), palmitic (C 16:0), and stearic acid (C 18:0) : their concentrations (%) decrease from AMF-41°C to AMF-20°C; whereas a concentration increase is observed for the major unsaturated acid : oleic acid (C 18:1). Major fatty acids in coconut oil are lauric acid (C 12:0) and myristic acid (C 14:0).

Table III. Calculated weighted average of carbon number and degree of unsaturation in commercial fats.

Fatty acids	AMF-41°C	AMF-32°C	AMF-20°C	Coconut oil
CN	15.03	14.79	14.75	12.10
DU	0.17	0.26	0.28	0.14

There are nearly no differences between weighted averages of degree of unsaturation for the different commercial fats, however there is a decrease of weighted averages of carbon number from AMF-41°C to coconut oil.

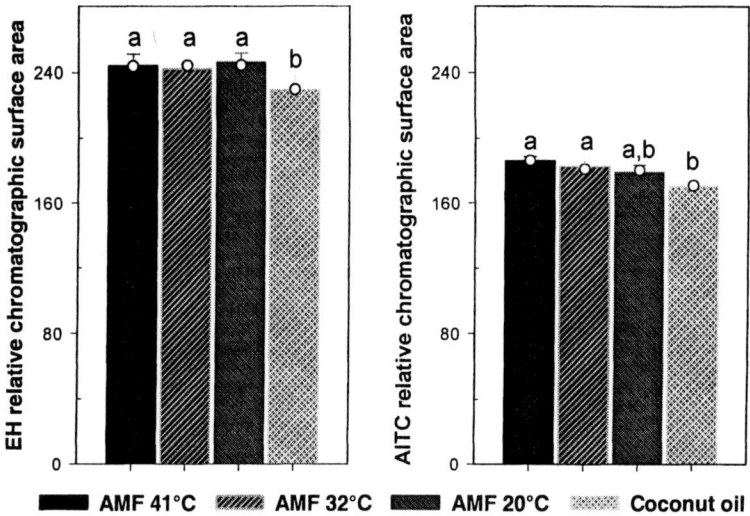

Figure 4. Ethyl hexanoate (left) and allyl isothiocyanate (right) chromatographic surface area in different commercial fats at 45°C. Bar charts represent experimental data; circles represent calculated data (Reproduced with permission from reference 5. Copyright 2003 Lavoisier SAS.)

There is a good agreement between experimental data and calculated data (Figure 4). This suggests that the model is suitable for predicting flavor release from fats differing in their fatty acids composition.

Second application : commercial oils with different compositions and corresponding emulsions

The fatty acid profile of the oils was determined to calculate the weighted average of carbon number and degree of unsaturation (Table IV).

Major fatty acids are unsaturated fatty acids with oleic acid, linoleic acid and linolenic acid. Oleic acid percentage decreases from olive oil to linseed oil, whereas linoleic acid content increases except for linseed oil, which contains more than 55 % of linolenic acid.

There are nearly no differences between weighted averages of carbon number for the different oils, however there is an increase of weighted averages of degree of unsaturation from olive oil to linseed oil.

Table IV. Oils composition: fatty acids profile (GC area %)

Fatty acids	Olive	Oleic sunflower	Rapeseed	Sunflower	Linseed
C 16:0	11.51	3.69	4.41	6.17	5.65
C 16:1	1.16	-	-	-	-
C 18:0	3.35	4.15	1.59	4.53	3.74
C 18:1	**76.36**	**77.69**	**59.76**	25.58	17.83
C 18:2	5.45	**12.00**	20.49	**61.67**	**16.71**
C 18:3	-	-	9.71	-	**55.11**
C 20:1	-	-	1.43	-	-
C 22:0	-	1.08	-	-	-

Table V. Oils calculated weighted average of carbon number and degree of unsaturation.

Fatty acids	Olive	Oleic sunflower	Rapeseed	Sunflower	Linseed
CN	0.88	1.02	1.31	1.49	2.17
DU	17.36	17.72	17.49	17.51	17.71

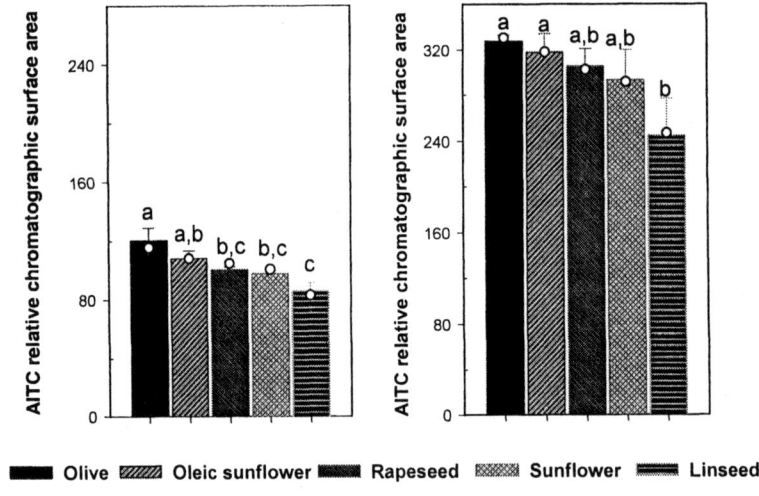

Figure 5. Allyl isothiocyanate chromatographic surface area in different pure oils (left) and emulsions (right) at 30°C. Bar charts represent experimental data; circles represent calculated data (Reproduced with permission from reference 5. Copyright 2003 Lavoisier SAS.)

There is a good agreement between experimental data and calculated data (Figure 5), both for pure oils and for emulsions. This suggests that in our experiment the main factor controlling volatility from these emulsions is the oil composition. Indeed the emulsion structure and the conformation of the protein at the interface affects flavor release (*10, 11*), then we have been tried to realize emulsions with the same droplet size. In our experiment, the differences in mean droplet sizes between the emulsions (9.9 μm ± 0.87 μm) are too low to induce significant differences in flavor release. This is in agreement with the results of Charles *et al.* (*9*) on the model systems.

Conclusions

The proposed linear model correlates flavor release with degree of unsaturation and carbon number for pure triacylglycerols. An increase in degree of unsaturation induces a decrease in flavor release, and an increase in chain length an increase in flavor release. The coefficients of the linear model follow an exponential law as a function of temperature. This model was successively applied to complex lipids and emulsions. Notice that further studies are needed to obtain a generalized model, in order to avoid analyzing separately the different sets of data. This predictable model could then be extended to other flavor compounds.

As a perspective, a quantitative structure property relationship (QSPR) study will complete the mathematical study, in order to correlate lipid properties with flavor release. Furthermore, the effect of the nature of lipids, and the effect of the solid to liquid ratio, will be both studied more intensively.

Acknowledgements

I would like to thank Dr. Anne TROMELIN (UMR Arômes - INRA Dijon) for her helpful discussion, Pierre JUANEDA (Unité de ntrition lipidique - INRA Dijon) for his knowledge about lipid characterization, and Hildegard IRL for her technical participation.

References

1. Guichard, E. *Food Rev. Int.* **2002**, *18*, 49-70.
2. Welsh, F. W.; Williams, R. E. *J. Chem. Technol. Biotechnol.* **1989**, *46*, 169-178.
3. Vedovati, C. Report, Conservatoire National des Arts et Métiers, Dijon, FR, 1995.

4. Maier, H. G.; In *Aroma research;* Maarse, H. and Groenen, P. J., Ed.; Pudoc : Wageningen, NL, 1975; pp 143-157.
5. Guichard, E.; Tromelin, A.; Juteau, A.; Rega, B.; Roudnitzky, N. In *Flavour research at the dawn of the twenty-first century;* Le Quéré, J. L. and Etiévant P.X., Ed.; Lavoisier: Paris, FR, 2003, pp 15-20.
6. Wolff, R.L. ; Fabien R. J. *Le Lait* **1989**, *69*, 33-46.
7. Morrisson, W. R.; Smith, L. M. *J. Lipid Res.* **1964**, *5*, 600-608.
8. Espinoza-Diaz, M. A. Ph.D. thesis, Université de Dijon, Dijon, FR, 1999.
9. Charles, M.; Lambert, S.; Brondeur, P.; Courthaudon, J. L.; Guichard, E. In *Flavor release;* Roberts, D. and Taylor, A. J., Ed.; American Chemical Society: Washington, DC, 2000; pp 342-354.
10. Charles, M.; Rosselin, V.; Beck, L. ; Sauvageot, F. ; Guichard, E. *J. Agric. Food Chem.* **2000**, *48*, 1810-1816.
11. Voilley, A.; Espinoza-Diaz, M.A.; Druaux, C.; Landy P. In *Flavor release;* Roberts, D. and Taylor, A. J., Ed.; American Chemical Society: Washington, DC, 2000; pp 142-152.

Chapter 16

Fatty Acid and Volatile Flavor Profiles of Textured Partially Defatted Peanut

Margaret J. Hinds[1], M. N. Riaz[2], D. Moe[3] and D. D. Scott[3]

[1]Nutritional Sciences Department, Oklahoma State University, Stillwater, OK 74078
[2]Food Protein Research and Development Center, Texas A&M University, College Station, TX 77843
[3]Food and Agricultural Products Research and Technology

Textured peanut (TP) may have potential as a meat substitute because its low level of volatile flavor compounds would give rise to an overall bland taste with no off-flavors. This study evaluated lipid and volatile flavor profiles of two sizes (5-6 and 10-14 mm) of TP (7.8% fat) prepared by twin-screw extrusion processing of partially-defatted peanuts, and held at 10-12°C for up to one year. Fatty acids were analyzed by GC. Static headspace volatiles were analyzed by GC-MS. The TP had a monounsaturated to saturated fatty acid ratio of 74:17. All 20 volatile flavor compounds in the TP were present at below-threshold levels, and were not affected by TP size. Levels of lipid oxidation products were not affected by storage, except for octanal and hexanoic acid which decreased with time. Hexanal was present at 0.1-0.2 µg/g indicating that TP would impart no grassy or beany flavor.

Textured peanut from extrusion processing

Extrusion processing is a process by which moistened starch and proteinaceous materials are plasticized by a combination of moisture, heat and mechanical shear *(1)*. High-shear extrusion processing imparts a textured or fibrous structure to proteinaceous materials of plant origin *(1)*. Studies to transform high-protein plant materials to textured products with meat-like properties started in the 1960s *(2)*, and have focused mainly on soy. Textured soy is commercially available, and is used extensively as a meat extender *(3)*, and for preparation of meat analogs *(4)*, but its inherent 'beany' flavor still remains a challenge *(5)*.

Since the mid-1970s, there have been sporadic studies utilizing both single-screw and twin-screw extruders to produce textured peanut *(6-10)*. These studies focused on attainment of optimal physical characteristics, especially structural integrity and meat-like properties, and protein content of the textured peanut. To date, no information on potential flavor of textured peanut has been reported. Typical extrusion processing temperatures range from 40 to 180°C, and can therefore facilitate lipid oxidation. Low-molecular weight carbonyl compounds, the typical end-products of lipid oxidation in oilseeds, may bind to the proteins and impart off-flavors *(11)*. Thus, it is important to investigate potential value-added utilization of textured peanut in terms of its flavor compounds.

Fatty acid profile and cardiovascular health

Studies have shown that fat profile is just as important as total fat in influencing cardiovascular health *(12,13)*. Low-fat diets in which lipid is replaced by carbohydrates raise triacylglycerols and decrease HDL-cholesterol, and may adversely affect cardiovascular health *(14)*. Diets containing lipids that are high in monounsaturated fat reduce the risk of cardiovascular disease by reducing the levels of total and LDL-cholesterol and triacylglycerols while maintaining HDL-cholesterol levels *(14)*. Oil from traditional low-oleic peanut cultivars contain 43-62% oleic acid, the major monounsaturated fatty acid *(15,16)*. Due to their high monounsaturated to saturated fatty acid content, peanuts can reduce risk of coronary heart disease *(14,17)*. Thus, it is important to evaluate the fatty acid profile of textured peanut to ascertain if extrusion processing has allowed it to retain the typical fatty acid profile of peanuts.

Volatile flavor compounds

Lipid oxidation is one of the major causes of deterioration of food because it leads to the development of off-flavors and off-odors, which can decrease acceptability. Rate of lipid oxidation is influenced by many parameters, including oxygen concentration, temperature, surface area, and water activity *(18)*. Hydroperoxides, the primary products of lipid oxidation are relatively unstable, and decompose into various compounds depending on the position of cleavage *(18)*. Important secondary products of lipid oxidation are aldehydes, aliphatic hydrocarbons, alcohols, carboxylic acids, ketones, and furans. For example, heptanal, octanal, nonanal, decanal, 2-decenal, and 2-undecenal are formed from hydroperoxides of oleic acid *(18,19,20)*. Hexanal, heptanal, 2-heptenal, octanal, 2-octenal, 2-decenal, 2,4-decadienal, 1-octen-3-ol, and 2-pentylfuran are produced from linoleic acid *(18,19,20)*. Propanal, 2-pentenal, 2,4-heptadienal, 3-hexenal, and 2,5-octadienal are from linolenic acid *(20)*. Many secondary lipid oxidation products are volatile.

The volatile flavor components of roasted peanuts have been extensively evaluated by numerous investigators *(21,22)*. They consist mainly of pyrazines, ketones, aldehydes, pyrroles, furans, terpenes, aromatic hydrocarbons, phenols, and alcohols.

Commercial devices for concentration of volatiles prior to GC analysis have been developed *(22)*. This has facilitated quantitation of lipid oxidation products as well as components of roasted peanut flavor. Numerous studies correlating volatile compounds from GC analysis of peanuts and peanut products with sensory perception have also been carried out *(19,21-25)*. Results from these studies have shown that each volatile compound imparts a typical flavor or odor. For example, the pyrazines impart roasted nutty flavors *(21,26)* while hexanal is associated with beany and grassy flavors *(23,24)*. Furthermore, aroma and flavor threshold levels for human perception of many volatile components of peanut flavor and lipid oxidation products have been established by correlating GC and sensory data *(21,24-27)*. GC identification and quantification of volatile components of textured peanut would provide important information regarding the potential flavors which it would impart to consumers.

Objectives of study

The objectives of this study were to evaluate the fatty acid and volatile flavor profiles by GC of two sizes of textured peanut (TP) to find out the potential flavors they might impart after one and 12 months storage at 10-12°C. Findings would provide implications for value-added utilization of textured peanut produced from partially-defatted peanuts.

Materials and Methods

Experimental design

The effects of a 2x2 factorial of TP diameter (10-14mm [TPL] and 5-6mm [TPS]) and storage period (1 and 12 months) on moisture content, fatty acid and GC volatile components were evaluated. Two process replications were carried out, and triplicate samples were taken for evaluating the dependent variables.

Preparation of textured peanut

Food grade split peanut kernels (Runner variety, Birdsong Peanut Company, Gorman, Texas) were used. Total moisture, water activity, and crude lipid (%, dry weight basis) of the spilts were $2.9\pm0.12\%$, 0.43 ± 0.001, and $43.6\pm3.89\%$, respectively. Splits were selected because they are typically used for oil production. They were blanched using a Mini-Dehuller (Nutana Machine Co., Saskatoon, Saskatchewan, Canada). Skins from the blanched peanuts were removed using a Zig Zag Kice aspirator (Kice Industries, Wichita, Kansas). The blanched peanuts were conditioned at 60-65°C in a steam-jacketed kettle (W.C. Smith and Sons, Inc., Philadelphia, Pennsylvania) with continuous stirring in order to prevent over cooking of the peanuts while arresting enzymatic activity. The conditioned peanuts were partially defatted using a Komet Screw Oil Expeller (IBG Monoforts, Mönchengladbach, Germany). The expeller barrel was heated to 150°C externally in order to get better oil release from the peanuts. The peanut press-cake obtained from this expeller was milled to 60 mesh using a hammer mill to produce meal. Total moisture, water activity, and crude lipid (%, dry weight basis) of the meal were $5.8\pm0.41\%$, 0.47 ± 0.007, and $10.4\pm0.20\%$, respectively.

The meal was processed using a Wenger TX-52 twin screw extruder (Wenger Manufacturing, Inc., Sabetha, Kansas) to produce textured peanut. Some features of this extruder which made it particularly suitable for preparing the textured peanut are: (i) a pre-conditioner (Wenger DDC) which allowed controlled pre-moistening of the peanut meal with water ahead of the extruder barrel for 40 minutes to enable moisture penetration into all particles to facilitate protein denaturation and starch gelatinization; (ii) a co-rotating, fully-intermeshing self-wiping screw; and (iii) the barrel segments each equipped with individually-controlled cavities for circulating thermal fluids for heating and/or cooling. After the peanut meal was conditioned with water, it was processed in the barrel at temperatures of 40-120C, with varying pressure in the seven barrel zones, and a pressure of 500 psi in zone #7. The textured materials were forced through a venture die situated at the distal end of the barrel. A size reduction machine (Comitrol 3600, Urschel Laboratories Inc., Valparaiso, Indiana) was used to randomly cut the textured peanut (TP) into 10-14 mm (TPL) or 5-6 mm (TPS) diameter pieces. The TPL and TPS were dried in a Wenger Continuous Drier (Wenger Manufacturing, Inc., Sabetha, Kansas) at 70°C for 6 min. After cooling, the TPL and TPS were packed in high-density polyethene bags, and held at 10-12°C for up to 12 months.

Evaluation of moisture content of textured peanut

Total moisture of TPL and TPS was measured using a Denver IR Moisture Analyzer (Denver Instrument, Denver, Colorado) at 75°C, set in the percent mode. Water activity was measured using a Rotronic Water Activity Meter (Model A2101, Rotronic Instrument Corp., Huntington, NY).

Evaluation of fatty acid profile of textured peanut

Samples were ground and their total crude fat quantified (petroleum ether extraction Method #920.85, AOAC) *(28)*. Fatty acid methyl esters were prepared from the crude fat using methanol and borontrifluoride catalyst (#Ce 1c-89, AOCS) *(29)*. Fatty acid methyl esters (FAMEs) were analyzed using an Agilent 6890 gas chromatograph (Agilent Technologies, Inc., Wilmington, Delaware). The injector temperature was 220°C, and a 1µL sample was injected at a 300:1 split ratio. The esters were separated on a Supelco (Bellefonte, Pennsylvania) SP 2380 fused silica capillary column (30m x 0.250mm I.D. x 0.20µm film thickness) with carrier gas helium flowing at a linear velocity of 35cm/s. The temperature of the flame ionization detector (FID) was 300°C. The GC oven's temperature was initially 50°C, and this was ramped at 15°C/min to 185°C and held for 8 min, then ramped at 15°C/min to 230°C to facilitate separation of the esters. The FAMEs were identified and quantified by external composite standards (Nu-Chek-Prep, Inc., Elysian, Minnesota).

Evaluation of volatile flavor compounds of textured peanut

Volatile flavor components were extracted using a static headspace method. To prepare samples for analysis, 15g of TPL or TPS were ground in 100mL jars for 15s using a Waring blender. An exact four-gram ground sub-sample was then deposited into a 22mL headspace vial, flushed with nitrogen for 20 sec, then capped and sealed. Volatile compounds were extracted by a Tekmar 7000/7050 static headspace autosampler (Tekmar-Dohrmann Co., Cincinnati, Ohio). In the autosampler, samples were heated at 140°C for 10 min after which they were automatically pressurized at 8psi for 0.25 min. One milliter of headspace was then transferred by the autosampler along a transfer line (140°C) to the GC-MS.

Volatile compounds were separated, identified and quantified by GC-MS (Agilent 6890-5973, Agilent Technologies, Inc., Wilmington, Delaware). The injector temperature of the GC was 220°C. One milliter headspace sample was injected with split ratio of 20:1, and separated by a J&W (J&W Scientific, Folsom, California) DB-5MS fused silica capillary column (30m x 0.25mm I.D. x 0.50μm film thickness) with helium traveling at a linear velocity of 36 cm/s. The initial GC oven temperature of 40°C was ramped at 2°C/min to 110°C and held for 10 min, then ramped at 2°C/min to 120°C and held for 5 min. The ionizing and electron multiplier voltages of the MSD were 70 eV and 1576.5 kV, respectively. The temperatures in the quadrupole and ion source were 150°C and 230°C, respectively. The scan range of the MSD (scan mode) was 35 – 400 m/z, and the detector time was 60 min. Eluting compounds were tentatively identified by their mass spectra using the NIST 98 library, then subsequently confirmed and quantified by corresponding external standard solutions of 5-5000 ppm in methanol. Standards were purchased from Sigma-Aldrich (Milwaukee, Wisconsin).

Data analysis

Data were analyzed using the SAS computer package (SAS PC version 8.1) *(30)*. ANOVA was used to evaluate the effects of diameter size of the TP and storage duration on moisture, fat content, and volatile flavor compounds. Differences between treatment means were verified for significance (α=0.05) by Duncan's Multiple Range Test.

Results and discussion

Moisture Content of Textured Peanut

Total moisture (%) and water activity of the textured peanut (TP) were significantly (p<0.01) affected by diameter of the TP and by length of storage, but not (p>0.05) by interaction between diameter and storage duration (ANOVA results). At the two storage periods, total moisture and water activity levels of TPL were significantly (α=0.05) higher than those of TPS (Table I). Moisture (%) and water activity ranged from 5.8 to 7.9 and 0.44 to 0.61, respectively. Both treatments lost moisture during storage, and after 12 months, the moisture (%) of the TPL was similar (p>0.05) to that of the 1-month old TPS (Table 1). Water activity of the TPL decreased from 0.61 to 0.53, and that of the TPS from 0.54 to 0.44.

Water activity (a_w) of the textured peanut is important because in lipid-containing foods, the rate of lipid oxidation is influenced by water activity *(18)*. Typically, lipid oxidation rate is low when a_w is 0.3 to 0.4, and high when it is <0.25 or 0.55 to 0.85 *(18,20)*. Peanut pastes containing 12% fat (on a dry weight basis) and stored at 24°C under various a_w regimes were observed to undergo highest lipid oxidation at a_w = 0.76 and minimal oxidation at a_w = 0.44 *(31)*. High-oleic peanuts stored at 25°C exhibited highest oxidation when held under 0.67 a_w, followed by 0.12, 0.52, 0.44, and 0.33 a_w, respectively *(32)*. The a_w values of the TP (Table I) indicate that the potential for oxidation of their lipids would decrease with storage (10-12°C). Freshly-extruded TPL should be dried to a lower a_w to hinder yeast/mold growth and lipid oxidation.

Table I. Moisture Content of Textured Peanut after Storage at 10-12°C

Textured Peanut Treatments	Moisture (%)	Water Activity
TPL-1	7.9±0.16 a	0.61±0.006 a
TPS-1	7.0±0.20 b	0.54±0.000 b
TPL-2	7.3±0.06 b	0.53±0.001 b
TPS-2	5.8±0.37 c	0.44±0.001 c

Means for the same parameter (column) followed by a different letter are significantly different (α=0.05), Duncan's Multiple Range Test.
TPL = textured peanut, 10-14 mm diameter. 1= after 1 mo, 2 = after 12 mo storage.
TPS = textured peanut, 5-6 mm diameter. 1= after 1 mo, 2 = after 12 mo storage.

Fatty Acid Profile of Textured Peanut

TPL and TPS contained 7.8±0.15 and 7.8±0.13 % crude lipid (%, dry weight basis), respectively. The major fatty acids present in the crude lipid of the textured peanut were oleic (C18:1, 70-71%), linoleic (C18:2, 7-8%), and palmitic (C16:0, 6%) (Table II). TPS contained significantly (α=0.05) more linoleic and linolenic acids than TPL (Table II). There were no significant differences in the levels of other fatty acids due to the size of the TP. The crude lipid of the textured peanut contained a high proportion of monounsaturated (~74%) to saturated fatty acids (~17%) (Table II), indicating its potential positive effect on cardiovascular health.

Table II. Fatty Acids (wt %) in Crude Lipid of Textured Peanut

Fatty Acid	TPS (5-6mm)	TPL (10-14mm)
C14:0	0.09 a	0.09 a
C16:0	5.82 a	6.35 a
C16:1	0.23 a	0.26 a
C18:0	3.01 a	3.35 a
C18:1	69.99 a	71.16 a
C18:2	8.35 a	6.84 b
C18:3	0.39 a	0.14 b
C20:0	1.36 a	1.35 a
C20:1	2.45 a	2.47 a
C22:0	3.52 a	3.40 a
C22:1	0.33 a	0.30 a
C24:0	3.00 a	2.86 a
C24:1	0.73 a	0.69 a
Total saturated	16.8 a	17.4 a
Total monounsaturated	73.7 a	74.9 a
Total polyunsaturated	8.7 a	7.0 b

Means for the same fatty acid (row) followed by a different letter are significantly different ($\alpha=0.05$), Duncan's Multiple Range Test.

Respective linoleic and linolenic acid levels (%, crude lipid weight) were similar ($p>0.05$) in the raw peanuts (18.2, 2.2) and partially-defatted peanut meal (18.1, 2.3) indicating that defatting of warm (60°C) peanuts by the expeller did not promote oxidation. However, respective linoleic and linolenic acid levels in TPS (8.3, 0.39) and TPL (6.8, 0.14) were different ($p<0.01$), suggesting that post-extrusion drying promoted oxidation of both linoleic and linolenic acids in TPL, and of linolenic acid in TPS. The moisture contents of pre-dried TPS and TPL were similar because extruded materials from the same run were randomly size-reduced to produce TPS and TPL, and both TPS and TPL were dried under identical conditions (70°C for 6 min). The lower levels of linoleic and linolenic acids in the dried TPL suggest that during the drying process, oxidation rate of these fatty acids was higher than in the TPS. The larger diameter of the TPL would have caused a slower rate of moisture loss and the corresponding maintenance of a high water activity (>0.61), which would promote oxidation

(18,20), and possible hydrolysis of these fatty acids *(18)*. Further work is necessary to confirm these observations.

Volatile Flavor Compounds in Textured Peanut

A total of twenty volatile compounds, including nine aldehydes and six pyrazines, were identified and quantified. These compounds were present in all treatments of the textured peanut, but their levels were generally low ranging from 0.02 to 3.3 µg/g. To facilitate the discussion, the compounds are grouped into 'heterocyclic' and 'lipid oxidation products'.

ANOVA

Levels of the volatile flavor compounds were not significantly ($p>0.05$) influenced by the diameters of the textured peanut materials (TPS and TPL) nor by interaction between TP diameter and duration of storage. However, as refrigerated (10-12°C) storage time was increased from one to 12 months, there were significant ($p<0.01$) decreases in quantities of the following compounds: two heterocyclic compounds, N-methylpyrrole and 2-ethyl-3-methylpyrazine; and two lipid oxidation products, hexanoic acid and octanal.

Heterocyclic Compounds

Eleven heterocyclic compounds were identified in the textured peanut (Table III). All of these compounds have previously been identified in roasted peanuts *(21)*, and this indicates that some degree of roasting of the peanut meal occurred during extrusion processing. Typical strong roasted peanut flavor tends to be detected when nuts are roasted at 175°C for 10 to 15 min *(26)*. In our study, temperatures in the extruder barrel zones ranged from 40 to 120°C, and this explains the low levels of these compounds in the TP.

N-methylpyrrole was present in the largest quantity, with levels of 3.3 µg/g in the 1-month old samples and 2.3-2.4 µg/g in the year-old samples (Table III). N-methylpyrrole is formed from the reaction between amino acids and sugars *(21)*, and is reported to impart a musty flavor *(23)* but its threshold level for perceptibility has not been reported. However, the results indicate that any potential musty flavor would decrease with storage of the TPS and TPL because levels of N-methylpyrrole decreased significantly ($p<0.05$) with storage (Table III).

Table III. Volatile Heterocyclic Compounds (μg/g) in Textured Peanut

Compound	TPS-1	TPL-1	TPS-2	TPL-2
N-methylpyrrole	3.3 a	3.3 a	2.4 b	2.3 b
methylpyrazine	0.04 a	0.04 a	0.02 a	0.02 a
2,5-dimethylpyrazine	0.8 a	0.8 a	0.6 a	0.5 a
2,6-dimethylpyrazine	0.03 a	0.03 a	0.03 a	0.03 a
2-ethyl 5-methylpyrazine	0.07 a	0.07 a	0.09 a	0.10 a
2-ethyl 3-methylpyrazine	0.3 a	0.3 a	0.02 b	0.02 b
phenylacetaldehyde	0.3 a	0.3 a	0.2 a	0.3 a
3-ethyl 2,5-dimethylpyrazine	0.1 a	0.1 a	0.1 a	0.1 a
toluene	0.1 a	0.1 a	0.09 a	0.06 a
benzaldehyde	0.04 a	0.04 a	0.04 a	0.03 a
2-furfural	0.02 a	0.02a	0.02a	0.02a

Means for the same compound (row) followed by a different letter are significantly different ($\alpha=0.05$), Duncan's Multiple Range Test.
TPL = textured peanut, 10-14 mm diameter; 1= after 1 mo, 2 = after 12 mo storage.
TPS = textured peanut, 5-6 mm diameter; 1= after 1 mo, 2 = after 12 mo storage.

Six alkylpyrazines were present in the textured peanut (Table III). These pyrazines are important products of Maillard reaction between amino acids and reducing sugars *(21)*, and impart predominantly roasted and/or nutty flavors *(21,24,26)*. In addition, grilled chicken, malty/chocolate, and slightly sweet aromas have been perceived from methylpyrazine, 2,5-dimethylpyrazine, and 3-ethyl-2,5-dimethylpyrazine, respectively *(24)*. Levels of methylpyrazine, 2,5-dimethylpyrazine, 2,6-dimethylpyrazine, 2-ethyl-5-methylpyrazine, and 3-ethyl-2,5-dimethylpyrazine were similar ($p>0.05$) in all treatments irrespective of TP size or storage period, and ranged from 0.02 to 0.8 μg/g (Table III). However, after one year, both TPS and TPL contained significantly ($\alpha=0.05$) lower levels of 2-ethyl-3-methylpyrazine (Table III) indicating a possible decrease in potential roasted nut flavor. Of all pyrazines present in roasted peanuts, 2,5-dimethylpyrazine is the most highly correlated to roasted peanut flavor and aroma, and can be used as an excellent sensory predictor *(26)*. The concentration of 2,5-dimethylpyrazine in roasted peanuts, and its odor detection in oil have been reported as 11 and 17 ppm, respectively *(21)*, and its aroma threshold in air as 50 ppm *(24)*. In roasted normal-oleic Florunner peanuts, minimal perceptible and strong roasted peanut flavors corresponded to 5.6 and 49.7 ppm, respectively *(26)*. In our study (Table III), 2,5-dimethylpyrazine levels in the TP were relatively low, and decreased from 0.8 μg/g (0.8 ppm) to 0.5-0.6 μg/g (0.5-0.6 ppm) after one-year storage indicating that the TP would not impart strong roasted peanut aroma or flavor.

All treatments of TP contained similar very low levels (0.02 μg/g) of 2-furfural (Table III), which is also a Maillard reaction product. It may impart sweet, woody, almond, baked bread and caramel flavors *(27)*. Benzaldehyde levels ranged from 0.03 to 0.04 μg/g (Table III). Its odor threshold level in water is 350 ppm, and it is reported to impart bitter almond, sweet, and aomatic flavors *(27)*. TP treatments contained 0.2-0.3 μg/g phenylacetaldehyde (Table III). This compound may impart floral, sweet, and caramel flavors, and has an aroma threshold of 10 ppm in air *(24)* and 4 ppm in water *(27)*. It also contributes to the 'bouquet' usually associated with warm, freshly-roasted peanuts *(21)*. Toluene levels were low (0.1 μg/g) in the one-month old TP, and decreased insignificantly (p>0.05) with storage (Table III). No threshold levels for perceptibility of toluene have been reported.

Lipid Oxidation Products

Nine lipid oxidation products, including six aldehydes, were identified in the textured peanut (Table IV). Hexanal, octanal, nonanal, and 2-pentylfuran are important contributors to the oxidized flavor of lipids *(18,19,24,25)*.

There were no significant (p>0.05) differences in hexanal levels (0.1-0.2 μg/g) due to storage of the TP (Table IV) whereas hexanal levels usually increase when peanut products are stored at 25°C or above *(19,24,25)*. This lack of increase in hexanal also indicates the relatively slow rate of lipid oxidation in the TP during storage as was expected from their water activity levels (Table I). Hexanal is considered as a major indicator of rancidity in peanuts and correlates positively with sensory perception *(19,24,25,33)*. It is reported to impart beany *(23)*, fatty, green, and grassy *(24,27)* flavors. Aroma thresholds for hexanal in air and water are 10 ppm *(24)* and 4.5 ppm *(27)*, respectively. Lee et al. *(33)* observed a significant correlation (r=0.77, p<0.001) between hexanal and sensory rancidity during 40-60°C storage of whey-protein-coated peanuts. Grosso and Resurreccion *(25)* reported that hexanal levels higher than 5.39 μg/g in cracker-coated peanuts and 7.40 μg/g in roasted peanuts would indicate unacceptability of these products to consumers. The relatively low levels of hexanal present in the TP even after one-year refrigerated storage indicate that TP would impart no off-flavors due to hexanal.

Table IV. Volatile Lipid Oxidation Products (µg/g) in Textured Peanut

Compound	TPS-1	TPL-1	TPS-2	TPL-2
2-propanone	1.8 a	1.8 a	1.4 a	1.4 a
2-methylpropanal	0.5 a	0.5 a	0.6 a	0.5 a
3-methylbutanal	0.7 a	0.7 a	0.9 a	0.8 a
2-methylbutanal	0.5 a	0.5 a	0.9 a	0.8 a
hexanal	0.1 a	0.1 a	0.2 a	0.2 a
hexanoic acid	1.4 a	1.4 a	0.02 b	0.02 b
nonanal	0.1 a	0.1 a	0.1 a	0.1 a
octanal	0.09 a	0.09 a	0.02 b	0.02 b
2-pentylfuran	0.05 a	0.05 a	0.06 a	0.04 a

Means for the same compound (row) followed by a different letter are significantly different ($\alpha=0.05$), Duncan's Multiple Range Test.
TPL = textured peanut, 10-14 mm diameter; 1= after 1 mo, 2 = after 12 mo storage.
TPS = textured peanut, 5-6 mm diameter; 1= after 1 mo, 2 = after 12 mo storage.

Octanal was the only aldehyde in the TP that changed with storage, and its levels decreased significantly ($p<0.05$) from 0.09 to 0.02 µg/g after one year (Table IV). Octanal is derived from both oleic and linoleic acids, and the crude lipid in the TP contained 70-71% oleic and 7-8% linoleic acids, respectively (Table II). The significant decrease in octanal definitely indicates that the rate of lipid oxidation in the TP decreased with storage. Octanal imparts fatty, citrus, and honey flavors and has a threshold of 0.5 ppm in water *(27)*. Thus, octanal levels in the TP would not contribute appreciably to its flavor.

TP contained 0.1 µg/g nonanal irrespective of size and storage period (Table IV). Nonanal is associated with floral *(24,27)*, citrus, orange, rose, fatty, and waxy *(27)* flavors, and has a threshold of 1 ppm in water *(27)*, and 10 ppm in air *(24)*. Levels of 2-pentylfuran in the TP were very low (0.05-0.06 µg/g) (Table IV). The compound 2-pentylfuran may impart beany, metallic, melon, and vegetable flavors, and has a threshold of 6 ppm in water *(27)*. It is one of the compounds responsible for reversion-flavor (beany, grassy) in soybean oil, and has also been observed to produce reversion-like flavors in other oils at levels of 2 ppm *(18)*. These results indicate that TP treatments would not impart off-flavors due to the quantities of nonanal, and 2-pentylfuran they contain.

The short-chain aldehydes, 2-methylpropanal, 2-methylbutanal, and 3-methylbutanal, had levels ranging from 0.5 to 0.9 µg/g, and these were not affected ($p>0.05$) by storage period (Table IV). Levels of these compounds typically increase when peanuts are roasted *(21)*. These compounds have been observed to impart dark roasted *(24)* and fruity flavors *(23,27)*. The threshold levels for 2-methylbutanal and 3-methylbutanal in water are 1.0 and 0.2 ppm, respectively, and they may also impart pungent odors *(27)*. After 1-year storage, TPS and TPL respectively contained 0.9 and 0.8 µg/g (ppm) of 2-methylbutanal, and might impart slight roasted, fruity or pungent odors because their 2-methylbutanal levels are close to its threshold.

Among the lipid oxidation products in the TP, 2-propanone was present in greatest abundance, and ranged from 1.4 to 1.8 µg/g (Table IV). It is associated with ethereal, apple, and pear flavors, and has an aroma threshold of 450,000 ppm in water *(27)*. This high threshold level indicates that the TP would not impart flavor notes arising from 2-propanone.

Hexanoic acid decreased ($p<0.05$) from 1.4 to 0.02 µg/g with storage time (Table IV). It may impart rancid, sour, sharp, cheesy, pungent and fatty flavors, and has a threshold of 3,000 ppm in water *(27)*. Hexanoic acid is formed from oxidation of hexanal. The relative hexanal and hexanoic acid levels in the TP suggest that the rate of hexanal oxidation was slower in one-year old TP treatments (Table IV), and can be explained by the decrease in water activity of the TP during storage.

Summary and Significance of Findings

The TP contained 7.8% crude lipid, which had a monounsaturated to saturated fatty acid ratio of 74:17, suggesting potential positive effects on cardiovascular health. There were no differences in volatile compounds due to TP size. The TP may lose some of its potential nutty and musty flavors after one-year storage due to decreases in 2-ethyl-3-methylpyrazine and N-methylpyrrole, respectively. Although water activity of the TPL was higher than that of the TPS, larger amounts of lipid oxidation products were not observed in the TPL. Water activity of the TPL and TPS decreased with storage, and respective values were 0.53 and 0.44 after one year indicating that lipid oxidation rate would be minimal in one-year old TP. Octanal and hexanoic acid were the only lipid oxidation products that changed significantly with storage, and they were observed to decrease, indicating that rate of lipid oxidation in the TP decreased with storage. Thus, after one-year, stored (10-12°C) TP would potentially impart even less fatty/rancid flavors than one-month old treatments. Very noticeable were the low levels of hexanal (0.1-0.2 µg/g or ppm) in the TP compared with reported threshold levels of 4.50-10 ppm indicating that TP definitely would not impart beany/grassy flavors, which are typically associated with textured soy. Levels of all volatile compounds in the TP were very low irrespective of diameter and storage period. They were below their reported threshold levels for human perception thereby indicating that the TP would impart negligible typical flavors associated with roasted peanuts and off-flavors arising from lipid oxidation. Thus, there is potential for value-added utilization of textured peanut size-reduced to either 10-14 or 5-6 mm diameter, and held in storage at 10-12°C for up to one year.

The findings of this study indicate that good quality textured peanut (in terms of fatty acid and volatile flavor profile) can be produced from partially-defatted peanut splits. They also suggest that the textured peanut would have a bland flavor, and this would facilitate its value-added utilization as a meat extender and/or meat analog.

References

1. Pham, C.B.; Del Rosario, R.R. *J. Food Technol.* **1984**, *19*, 535-547.
2. Sheard, P.R.; Ledward, D.A.; Mitchell, J.R.; *J. Food Technol.* **1985**, *19*, 475-483.
3. Hettiarchchy, N.; Kalapathy, U. In *Soybeans: Chemistry, Technology, and Utilization*; Liu, K., Ed.; Chapman and Hall: New York, NY, 1997, pp 379-411.
4. Buffett, H.G. *INFORM* 1994, *5*, 1156-1158.
5. Pehanich, M. *Wellness Foods.* May/June 2002, p 18, 19, 22-23.
6. Ayres, J.L.; Branscomb, L.L.; Rogers, G.M. *J. Am. Oil Chem. Soc.* **1974**, *51*, 133-135.
7. Aguilera, J.M.; Rossi, F.; Hiche, E.; Chichester, C.O. *J. Food Sci.* **1980**, *45*, 246-250, 251.
8. Alid, G.; Yanez, E.; Aguilera, J.M.; Monckeberh, F.; Chichester, C.O. *J. Food Sci.* **1981**, *46*, 948-949.
9. Aboagye, Y.; Stanley, D.W. *Can. Inst. Food Sci. Technol. J.* **1987**, *20*, 148-153.
10. Hinds, M.J.; Phillips, R.D. *Proc. Am. Peanut Res. Educ. Soc.* 1999, *31*, 61.
11. Damodaran, S. In *Food Chemistry*; Fennema, O.R., Ed.; Marcel Dekker, Inc.: New York, NY, 1996; pp 321-429.
12. Hu, F.B. *New Engl. J. Med.* **1997**, *337*, 1491-1499.
13. Kris-Etherton, P.M.; Yu, S. *Am. J. Clin. Nutr.* **1997**, *65*, 1628S-1644S.
14. Kris-Etherton, P.M.; Pearson, T.A.; Wan,Y.; Hargrove, R.L.; Moriarty, V.F.; Etherton, T.D. *Am. J. Clin. Nutr.* **1999**, *70*, 1009-1015.
15. Brown, D.F.; Carter, C.M.; Mattil, K.F.; Darroch, J.G. *J. Food Sci.* **1975**, *40*, 1055-1057.
16. Hinds, M.J. *Food Chem.* **1995**, *53*, 7-14.
17. Hu, F.B.; Stampfer, M.J.; Manson, J.E.; Rimm, E.B.; Colditz, G.A.; Speizer, F.E.; Hennekens, C.H.; Willett, W.C. *Brit. Med. J.* **1998**, *317*, 1341-1345.
18. Nawar, W.W. In *Food Chemistry*; Fennema, O.R., Ed.; Marcel Dekker, Inc.: New York, NY, 1996; pp 225-319.
19. Bett, K.L.; Boylston, T.D. In *Lipid Oxidation in Food*; St. Angelo, A.J., Ed.; ACS Symposium Series 500, American Chemical Society: Washington, DC, 1992; pp 322-350.

20. deMan, J.M. *Principles of Food Chemistry*; Aspen Publishers, Inc.: Gaithersburg, MD, 1999; pp 1-110.
21. Ahmed, E.M.; Young, C.T. In *Peanut Science and Technology*; Pattee, H.E.; Young, C.T. Eds.; American Peanut Research and Education Society, Inc.: Yoakum, TX, 1982; pp 655-688.
22. Vercellotti, J.R.; Mills, O.E.; Bett, K.L.; Sullen, D.L. In *Lipid Oxidation in Food*; St. Angelo, A.J., Ed.; ACS Symposium Series 500, American Chemical Society: Washington, DC, 1992; pp 232-265.
23. Young, C. T., Hovis, A.R. *J. Food Sci.* **1990**, *55*, 279-280.
24. Braddock, J.C.; Sims, C.A.; O'Keefe, S.F. *J. Food Sci.* **1995**, *60*, 489-493.
25. Grosso, N.R.; Resurreccion, A.V.A. *J. Food Sci.* **2002**, *67*, 1530-1537.
26. Baker, G.L.; Cornell, J.A.; Gorbet, D.W.; O'Keefe, S.F.; Sims, C.A.; Talcott, S.T. *J. Food Sci.* **2003**, *68*, 394-399.
27. FlavorWORKS. *Flavor and Fragrance Database*; Version 2.01, 2001; Flavometrics, Inc.: Anahaeim, CA.
28. AOAC. *Official Methods of Analysis of AOAC International, 17th Ed.;* Horwitz, W., Ed.; Association of Official Analytical Chemists: Washington, D.C., 2000.
29. AOCS. *Official Methods and Recommended Practices of the American Oil Chemists Society*, AOCS: Champaign, IL, 1997.
30. SAS. Statistical Applied Systems; SAS Institute, Cary, NC, 2000.
31. Hinds, M.J. *Proc. Am. Peanut Res. Educ. Soc.* **1995**, *27*, 46.
32. Baker, G.L.; Sims, C.A.; Gorbet, D.A.; Sanders, T.H.; O'Keefe, S.F. *J. Food Sci.* **2002**, *67*, 1600-1603.
33. Lee, S.-Y.; Trezza, T.A.; Guinard, J.-X.; Krochta, J.M. *J. Food Sci.* **2002**, *67*, 1212-1218.

Author Index

Blank, I., 19
Boelrijk, Alexandra E. M., 49
Burgering, Maurits J. M., 49
de Jongh, Harmen, 87
de Roos, Kris B., 145
de Wijk, R. A., 95, 105
Elmore, J. Stephen, 35
Enser, Michael, 35
Fabre, M., 61
Guichard, Elisabeth, 61, 191
Hamam, Fayez, 3
Hinds, Margaret J., 205
Janssen, Anke, 87
Jellema, R. H., 105
Khan, M. Ahmad, 3
Leser, M. E., 19
Lin, J., 19
Linforth, R. S. T., 159
Linssen, Jozef P. H., 49
Löliger, J., 19
Moe, D., 205
Mottram, Donald S., 35
Prinz, J. F., 95
Relkin, P., 61
Riaz, M. N., 205
Rondaut, Gaëlle, 191
Rota, Valerie, 73
Roudnitzky, Natacha, 191
Schieberle, Peter, 73
Scott, D. D., 205
Shahidi, Fereidoon, 3
Taylor, A. J., 159
Tunick, Michael H., 133
van den Oever, G. J., 171
Van Hekken, Diane L., 133
van Loon, Wil A. M., 49
Voragen, Alphons G. J., 49
Weenen, Hugo, 87, 95, 105, 119
Wood, Jeffrey D., 35

Subject Index

A

AA. *See* Arachidonic acid
Acarbose inhibition of α-amylase, 126–127, 128*f*
2-acetyl-1-pyrroline, 79
ALA. *See* Alpha-linolenic acid
Algal oils
  acidolysis, 8
  ARASCO (arachidonic acid single cell oil) stability, 17*t*, 18
  DHASCO (docosahexaenoic acid single cell oil) stability, 17*t*, 18
  effect of chlorophyll on stability, 13
  effect of processing on stability, 13, 17*t*
  Omega Gold oil stability, 17*t*, 18
Allyl isothiocyanate
  chromatographic surface area in different commercial fats, 201*f*
  effect of temperature on aroma volatility, 196, 197*f*
  emulsion preparation, 194
  static headspace analysis, 194–195
Alpha-linolenic acid (ALA, C18:3ω3), 7, 36
  See also n-3 fatty acids; Polyunsaturated fatty acids
2-Aminoacetophenone, 79
Amonton's law, 96
α-Amylase
  creamy mouthfeel and starch breakdown by α-amylase, 100–101, 121–122, 126–127, 128, 130
  inhibition by acarbose, 126–127, 128*f*
Anethole, 161–162, 164
Animal fat, 67, 68*t*, 69
ARA. *See* Arachidonic acid
Arachidonic acid (AA, ARA, C20:4ω6), 7
  See also n-6 fatty acids; Polyunsaturated fatty acids
Aroma Extraction Dilution Analysis (AEDA), 50, 74, 75–78
Aroma perception, 150–152
Aroma perception, effect of lipids, 152
ATR-IR. *See* Attenuated total reflection infrared (ATR-IR) spectra
Attenuated total reflection infrared (ATR-IR) spectra
  composition of oral coating, 91–94
  egg albumin (protein), 90*f*, 92*f*
  mayonnaise, 87, 89–94
  sunflower oil, 90*f*, 92*f*
  typical infrared vibrations, 91*t*
  xanthan, 90*f*, 92*f*
Δ^5-Avenasterol, 11

B

Beef flavor. *See* Grilled beef
Borage oil, 13, 16*f*
Brassicasterol, 11
Brick cheese, changes during ripening, 138, 139*f*
Business model relating food properties to consumer choice
  consumer choice level (level 5), 172, 174*f*
  consumer preference level (level 4), 172, 174*f*
  illustration, 174*f*
  level in the business model, explanation, 172, 174*f*
  microstructure control of properties in model, 172–173

multi-modal interactions, 173
oral physical properties level (level 2), 172, 174*f*
physiological variability, 173
pre-oral product properties level (level 1), 172, 174*f*
sensory level (level 3), 172, 174*f*
See also Integrated sensory (or consumer) response modelling

C

Canola (rapeseed) oil, 17*t*
Carotenes (α and β), 12
Carotenoids, 12, 13
Cattle breeds
  Aberdeen Angus, 36
  effect on fatty acids, 43, 45*t*, 46*t*
  effect on flavor, 36, 39, 41*t*, 42*t*
  Holstein-Friesian, 36
  Welsh Black, 36
Cattle diets
  effect of diet on aroma compounds in grilled beef, 39, 41*t*, 42*t*, 43
  effect of diet on fatty acids, 43, 45*t*, 46*t*
  n-6 fatty acids in concentrates-based diets, 36
  n-3 fatty acids in forage diets, 36
  phytol derivatives in silage-fed animals, 36, 39
Cephalin. See Phosphatidylethanolamine
Cheddar cheese, changes during ripening, 138, 139*f*
Cheese chemistry and rheology
  Brick cheese, changes during ripening, 138, 139*f*
  casein, proteolytic breakdown, 135, 139
  casein, types, 134–135
  Cheddar cheese, changes during ripening, 138, 139*f*
  chymosin, 135

Colby cheese, changes during ripening, 138, 139*f*
dynamic measurements, 138
empirical measurements, 137
factors affecting, 135
fundamental measurements, 137
Gouda cheese, changes during ripening, 138, 139*f*
Havarti, changes during ripening, 138, 139*f*
imitative measurements, 137
lipid content of cheese, 134
lipids, role in cheese flavor, 134
low-fat Mozzarella, development and properties, 139–140
Mozzarella cheese, changes during ripening, 135, 136*f*, 138, 139–140
Old Amsterdam cheese, changes during ripening, 138, 139*f*
production processes, 134
protein content, 134–135
ripening, 135, 136*f*
Romano cheese, changes during ripening, 138, 139*f*
stress-strain curves, 137
terminology, 137–138
texture map of various cheeses during aging, 138, 139*f*
texture maps, 138, 139*f*
texture profile analysis (TPA), 137, 140
torsion tests, 138
transient tests, 138
uniaxial compression, 137
Chewiness, definition, 137
Chlorophylls, 12, 13, 16*f*
Citrostadienol, 11
Coconut oil, 200*t*
Coenzyme Q. See Ubiquinones
Cohesiveness, definition, 137
Colby cheese, changes during ripening, 138, 139*f*
Confocal scanning laser microscopy (CSLM), 123–124, 126
Corn oil, 17*t*

Coulomb's law, 96
Creamy mouthfeel
  composition of custard desserts tested, 107, 108*t*, 109*t*
  correlation with texture attributes, 106, 112–113, 127–128, 129*f*
  effect of airy attribute, 88, 113, 115, 121
  effect of bitter chemical flavor, 106
  effect of dry mouthfeel, 113, 115
  effect of fatty attribute, 88, 113, 114, 121
  effect of granularity or grittiness, 106, 120, 128
  effect of oil droplet size, 110*t*, 120
  effect of oral processing, 124–125
  effect of rough mouthfeel, 106, 113, 114–115, 121, 127–128, 130
  effect of sickly flavor, 106
  effect of slippery attribute, 120
  effect of smooth attribute, 106, 120, 121
  effect of starch breakdown by α-amylase in saliva, 100–101, 121–122, 126–127, 128, 130
  effect of thick attribute, 106, 114, 120, 121, 128
  fat surfacing, 121–122, 124–127, 128
  partial least squares (PLS) analysis, 106, 107, 111–115, 116*t*
  PLS regression for creamy mouthfeel *versus* texture attributes, 113, 120–121
  sensory evaluation methods, 122–123
CSLM. *See* Confocal scanning laser microscopy (CSLM)
Custards
  composition of commercial custard desserts, 107, 109*t*
  composition of experimental custard desserts, 107, 109*t*
  evaluation by panel, 107, 122–123
  microstructural analysis by confocal scanning laser microscopy (CSLM), 123–124, 126
  PLS model for creamy mouthfeel, 107, 111–115
  relationship of friction to fat content, 99*f*, 100–101
  sensory attributes, 107

D

(E,E)-2,4-Decadienal
  formation from linoleic acid, 79*f*
  formation from phospholipids, 20, 24–27, 30
  in sheep meat, 76, 77, 78, 79–80, 81–82
(E)-2-Decenal, 20, 81–82
(E)/(Z)-2-Decenal, 24–27
DHA. *See* Docosahexaenoic acid
Diacetyl, 64, 65*f*, 65*t*, 69, 147*f*
Differential scanning calorimetry (DSC)
  measurement of melting behavior of fats, 64, 67, 69, 70*f*
  overview, 193, 195, 196*f*
Diffusion coefficient (D; Stokes-Einstein equation), 149–150
Docosahexaenoic acid (DHA, 22:6ω3), 7
DSC. *See* Differential scanning calorimetry
Dynamic headspace system model, 166–169

E

Edible oils
  fatty acid composition, 5–7
  non-triacylglycerols in, 8–13
  stability, 7–13
  triacylglycerols in, 5–7

Eicosanoids, 5
Eicosapentaenoic acid (EPA, 20:5ω3), 5, 7
  See also n-3 fatty acids; Polyunsaturated fatty acids
Emulsions
  oil-water partition coefficients, effect of chain length and saturation, 62, 148, 192
  sensory attributes of emulsified foods, 88
  texture characteristics, categories, 88
  volatility in (P_{ap}), 147
  See also Oil-in-water emulsions; Semi-solid foods
EPA. *See* Eicosapentaenoic acid
trans-4,5-epoxy-(E)-2-decenal, 24–26
4,5-Epoxy-(E)-2-decenal, 76, 77, 78, 81–82
Ethyl hexanoate
  chromatographic surface area in different commercial fats, 201*f*
  effect of temperature on aroma volatility, 196
  emulsion preparation, 194
  properties in emulsion, 64, 65*t*, 66*f*, 69, 71, 165–166
  static headspace analysis, 194–195
4-Ethyloctanoic acid, 74, 76, 77, 78, 79–80, 81
Evening primrose oil, 13, 14*f*–16*f*
Extrusion processing, 206

F

Fat reduction in foods, 160
  See also Business model relating food properties to consumer choice; Integrated sensory (or consumer) response modelling
Fat surfacing, 121–122, 124–126
Fatty acids
  branched chain fatty acids in sheep meat, 74, 80–81

commercial fats composition, fatty acids profile, 200*t*
composition of edible oils, 5–7
effect on odorant formation, 29–31
fatty acid profile and cardiovascular health, 206
weighted average of carbon number and degree of unsaturation in commercial fats, 200*t*
  See also Monounsaturated fatty acids; Polyunsaturated fatty acids (PUFA); Saturated lipids; specific compounds
Flavonoids, 12
Flavor, definition, 145
Flavor dilution (FD) factors of odorants, 23–24, 25*t*, 78*t*
Flavor perception
  effect of lipids on mass transport, 148–154
  effect of lipids on mouthfeel, 154–155
  effect of lipids on orthonasal aroma perception, 152
  effect of lipids on phase partitioning, 146–148
  effect of lipids on retronasal aroma perception, 150–152
  effect of lipids on taste perception, 154
  effect of lipids on total flavor, 155–156
  interactions between aroma, taste, mouthfeel, and overall flavor perception, 155*f*
Flavor release
  air-water partition coefficient (K_{aw}), 167, 168*f*
  balance of flavor compounds from low fat foods, 160
  breath volatile profiles for ethyl octanoate in water or lipid, 163–165
  diacetyl, properties in emulsion, 64, 65*f*, 65*t*, 69

dynamic headspace system model, 166–169
effect of dilution by saliva or water, 160, 163–165, 169
effect of hydrophobicity on flavor release in emulsions, 62, 67, 153f, 192
effect of lipids on flavor delivery in yogurt, 161–162, 163f, 164
effect of lipids on intensity of flavor delivery, 161–162, 165–166
effect of lipids on mass transfer, 160, 166–169
effect of lipids on persistence of flavor, 160, 162, 163–165, 167
effect of lipids on product/air partition coefficient, 160
effect of melting behavior of fats, 64, 67, 69, 70f
effect of vapor-oil partition coefficient (K_i), 63–64, 67, 68t, 69
effect physical state of fat, 62
ethyl hexanoate, properties in emulsion, 64, 65t, 66f, 69, 71, 165–166
French fries volatile compounds observed by MS-NOSE, 52, 53t
gas phase analysis methods, 160–161
hexenol, properties in emulsion, 65t, 66f, 67, 69
in vitro release in mouth model system, 51
maximum intensity (I_{max}), effect of frying time, 52–53, 54f
maximum intensity (I_{max}), effect of salt addition, 56, 57f
measurement of French fries flavor release by MS-NOSE, 50–51
measurement of released flavors from French fries, 51
solid-phase microextraction (SPME) measurement of aroma release, 63, 64, 65f–66f, 67

temperature, effect on partition between lipid and vapor phases, 62, 148
time of maximum intensity (t_{max}), effect of frying time, 53f, 55f
time of maximum intensity (t_{max}), effect of salt addition, 56, 58f
French fries
compounds observed by MS-NOSE, 52, 53t
cooking method, 50
in vitro flavor release in mouth model system, 51
maximum intensity of flavor release (I_{max}), effect of frying time, 52–53, 54f
maximum intensity of flavor release (I_{max}), effect of salt addition, 56, 57f
measurement of flavor release by MS-NOSE, 50–51
measurement of released flavors, 51
time of maximum intensity of flavor release (t_{max}), effect of frying time, 53f, 55f
time of maximum intensity of flavor release (t_{max}), effect of salt addition, 56, 58f
Friction
Amonton's law, 96
as system property, 96, 97
coefficient of friction, definition, 96
Coulomb's law, 96
definition, 96
kinetic friction, definition, 96
oral mucosa, coefficient of friction, 96
relationship to fat content of foods, 99f, 100–101
relationship to food attributes, 96, 100f, 101
relationship to oil droplet size, 101, 102f
saliva, role in lubrication, 96, 97, 100–101, 166

static friction, definition, 96
Fucosterol, 11

G

Gamma-linolenic acid (GLA, C18:3ω6), 5, 7
  See also n-3 fatty acids; Polyunsaturated fatty acids
GLA. *See* Gamma-linolenic acid
Gouda cheese, changes during ripening, 138, 139*f*
Green tea extracts, 7
Grilled beef
  analysis of fatty acids, 38
  analysis of volatile compounds, 37–38
  comparison of odor-active oxo-compounds in different types of meat, 80
  dimethyltrisulfide and dimethyl trisulfide in, 43, 44*t*
  effect of breed on aroma compounds, 39, 41*t*, 42*t*
  effect of breed on fatty acids, 43, 45*t*, 46*t*
  effect of diet on aroma compounds, 39, 41*t*, 42*t*, 43
  effect of diet on fatty acids, 43, 45*t*, 46*t*
  effect of slaughter age on aroma compounds, 39, 41*t*, 42*t*, 43, 44*t*
  effect of slaughter age on fatty acids, 45*t*, 46*t*, 47
  oxidation products of C18:2 n-6 in grilled beef, 39
  oxidation products of C18:3 n-3 in grilled beef, 39
  phytol derivatives in silage-fed animals, 36, 39
  volatile compounds in subcutaneous fat, 36
Guar thickening, texture and taste effects, 176*f*

H

Hardness, definition, 137
Havarti, changes during ripening, 138, 139*f*
Hexanal, 20, 207
Hexenol, properties in emulsion, 65*t*, 66*f*, 67, 69
3-Hydroxy-4,5-dimethyl-2(5H) furanone, 76
4-Hydroxy-2,5-dimethyl-3(2H) furanone, 76, 77*f*, 79
Hydroxytyrosol, 12

I

Integrated sensory (or consumer) response modelling (ISRM) (mayonnaise)
  coefficients of sensory attributes in bread liking model, 177*t*
  ISRM1, identification of sensory effects of fat replacement which have consumer relevance, 175–177
  ISRM2, qualitative analysis of the physics of the oral processes during consumption, 178–180
  ISRM3, identify/define food properties related to mayonnaise consumption, 181
  ISRM4, measure physical properties identify/define food properties related to mayonnaise consumption, 181
  ISRM5, measurement of wide range of products within category, 181
  ISRM6, quantitative models of sensory of consumer preference data, 186
  ISRM7, validation and finetuning of mechanistic understanding and microstructure control, 186, 188
  overview, 172, 174–175, 188–189

physical and consumer preferences
vs. fat level, 182f–183f
physical and sensory fattiness data
vs. fat level, 184f–185f
physical processes during
consumption of spread/bread
system, 178–180
quantitative model of drivers of
sensory fattiness of spread/bread
system, 187f
steps of ISRM method, 175
See also Business model relating
food properties to consumer
choice
ISRM. *See* Integrated sensory (or
consumer) response modelling

L

LA. *See* Linoleic acid
Lamb. *See* Sheep meat
Lecithin. *See* Phosphatidylcholine
Linoleic acid (LA, C18:2ω6)
in concentrates-based cattle diets, 36
oxidation products of C18:2 n-6 in
grilled beef, 39
products of linoleic acid oxidation,
207
stability, 7
See also n-6 fatty acids;
Polyunsaturated fatty acids
Linolenic acid (LNA, C18:3ω3)
α-linolenic acid in forage, 36
oxidation products of C18:3 n-3 in
grilled beef, 39
products of linolenic acid oxidation,
207
See also Alpha-linolenic acid (ALA,
C18:3ω3); Gamma-linolenic acid
(GLA, C18:3ω6); n-3 fatty acids;
Polyunsaturated fatty acids
Linseed oil, 201, 202t
Lipid oxidation products, 207
LNA. *See* Linolenic acid

Lubrication
measurement apparatus, 97, 98f
of bread crumbs by spreads, 175–176
saliva, role in lubrication, 96, 97, 100–101, 166
Lutein, 12

M

Maillard reaction, 50, 52, 214
Mass transfer, 160, 166–169
Mass transport
concentration gradients of aroma
compounds in water and air, 148–149
concentration gradients of aroma
compounds in water and
emulsions, 150–151
diffusion coefficient (D; Stokes-
Einstein equation), 149–150
effect of lipids on mass transport,
148–154
effect of lipids on orthonasal aroma
perception, 152
effect of lipids on retronasal aroma
perception, 150–152
effect of lipids on time-intensity
profile of aroma release, 153–154
effect of oil droplet size, 151
mass flux (J), 148–149
mass transport coefficient (k_p), 149
Mayonnaise
attenuated total reflection infrared
(ATR-IR) spectra, 87, 89–94
composition of samples, 88–89
oral coating composition, 91–94
physical and consumer preferences
vs. fat level, 182f–183f
physical and sensory fattiness data
vs. fat level, 184f–185f
quantitative model of drivers of
sensory fattiness of spread/bread
system, 187f

relationship of friction to fat content, 99f, 100–101
relationship of friction to oil droplet size, 101, 102f
See also Integrated sensory (or consumer) response modelling (ISRM) (mayonnaise)
4-Methylnonanoic acid, 74, 79–80
4-Methyloctanoic acid, 74, 81
Monounsaturated fatty acids, 5, 206
Mouthfeel, 146, 154–155
See also Creamy mouthfeel; Rough mouthfeel
Mozzarella cheese, changes during ripening, 135, 136f, 139–140
MS-NOSE, 50–51, 52, 53t
Multivariate data analysis. See Partial least squares (PLS) analysis
Mutton. See Sheep meat
Myristic acid (C14:0), 5

N

n-3 fatty acids (omega-3 fatty acids)
in forage diets of cattle, 36
oxidation products of C18:3 n-3 in grilled beef, 39
ratio of ω6 to ω3 fatty acids in western diet, 5
symbols used (ω3, n-3), 5
See also Polyunsaturated fatty acids (PUFA)
n-6 fatty acids (omega-6 fatty acids)
in concentrates-based cattle diets, 36
oxidation products of C18:2 n-6 in grilled beef, 39
ratio of ω6 to ω3 fatty acids in western diet, 5
symbols used (ω6, n-6), 5
See also Polyunsaturated fatty acids (PUFA)
National School Lunch Program, 139
Non-triacylglycerols (NTAG) in edible oils
characteristics, 4–5
content in edible oils, 8–13
effects of processing, 4–5, 8, 13–18
sources, 4
uses, 10t
(E,Z)-2,6-Nonadienal, 76, 77f, 79f
Nonanal, 20
(E)-2-Nonenal, 76, 77f, 79–80, 81–82

O

OA. See Oleic acid
Oat oil, 12
(Z)-1,5-Octadien-3-one, 76, 77
1-Octen-3-one, 20, 24–27, 76, 77f
2-Octenal, 24–27
Odor, definition, 145
Odor-active compounds and odorants
comparison of odor-active oxo-compounds in different types of meat, 80
concentration gradients of aroma compounds in water and air, 148–149
(E,E)-2,4-decadienal from phospholipids, 20, 24–27, 30
(E,E)-2,4-decadienal in sheep meat, 76, 77, 78, 79–80, 81–82
(E)-2-decenal, 20
(E)/(Z)-2-decenal, 24–27
(E)-2-decenal in sheep meat, 81–82
effects of lipids on volatility (P_{ap}) of aroma compounds, 147f
trans-4,5-epoxy-(E)-2-decenal, 24–26
4,5-epoxy-(E)-2-decenal in sheep meat, 76, 77, 78, 81–82
4-ethyloctanoic acid in sheep meat, 76, 77, 78, 79–80, 81
fatty acids, effect on odorant formation, 29–31
flavor dilution (FD) chromatogram of odorants from cooked sheep meat, 75, 76f

flavor dilution (FD) factors of
  odorants from phospholipids, 23–24, 25t
formation from phospholipids, 25–28
hexanal, 20
3-hydroxy-4,5-dimethyl-2(5H)
  furanone in sheep meat, 76
4-hydroxy-2,5-dimethyl-3(2H)
  furanone in sheep meat, 76, 77f, 79
4-methylnonanoic acid in sheep
  meat, 74, 79–80
4-methyloctanoic acid in sheep meat, 74, 81
model system for generation of odor-active compounds, 21–22, 28–29
(E,Z)-2,6-nonadienal sheep meat, 76, 77f, 79f
nonanal, 20
(E)-2-nonenalin sheep meat, 76
(Z)-1,5-octadien-3-one in, 76, 77
1-octen-3-one, 20, 24–27
1-octen-3-one in sheep meat, 76, 77f
2-octenal, 24–27
odor activity values (OAV), 27
odorants in aroma extract of
  phospholipids, 23–24, 25t
2-pentylpyridine in sheep meat, 74
polar moiety, effect on odorant
  formation, 31
precursors of odorants, 28
structures, 26f, 77f
(E,Z,Z)-2,4,7-tridecatrienal, 24–27, 29
(E)-2-undecenal, 24–27
volatile compounds in heated
  phospholipids, 20
Odor activity values (OAV), 27
Oil-in-water emulsions
apolarity of animal fat, 67, 69
diacetyl, properties in emulsion, 64, 65f, 65t, 69
differential scanning calorimetry
  (DSC) measurement of melting
  behavior of fats, 64, 67, 69, 70f
effect of hydrophobicity on flavor
  release, 62, 67, 153f, 192
effect physical state of fat on
  volatility of flavor compounds, 62
ethyl hexanoate, properties in
  emulsion, 64, 65t, 66f, 69, 71
food model emulsions, preparation, 62–63
hexenol, properties in emulsion, 65t, 66f, 67, 69
log P, effect on solubility, 62
low vapor pressure of nonpolar
  molecules dissolved in oil, 62
oil-water partition coefficients, effect
  of chain length and saturation, 62, 148, 192
solid-phase microextraction (SPME)
  measurement of aroma release, 63, 64, 65f–66f, 67
temperature, effect on partition
  between lipid and vapor phases, 62, 148
triacylglycerols, amounts in animal
  fat, 67, 68t, 69
vapor-oil partition coefficient (K_i), 63–64, 67, 68t, 69
See also Emulsions
Oilseeds, 4, 206
See also specific oils
Old Amsterdam cheese, changes
  during ripening, 138, 139f
Oleic acid (OA, C18:1ω9), 7, 207
See also Monounsaturated fatty acids
Oleic sunflower oil, 201, 202t
Olive oil
chlorophylls in, 12
effect of processing on stability, 13, 16f
fatty acid composition, 201, 202t
phenolics in olive oil, 12
squalene in, 12
weighted average of carbon number
  and degree of unsaturation, 202t

Omega-3 (ω-3) fatty acids. *See* n-3 fatty acids; Polyunsaturated fatty acids (PUFA); specific types
Omega-6 (ω-6) fatty acids. *See* n-6 fatty acids; Polyunsaturated fatty acids (PUFA); specific types
Oral coatings, 88, 91–94

P

Palm oil, 5
Partial least squares (PLS) analysis
 creamy mouthfeel, 106, 107, 111–115, 116t
 overview, 106, 107, 111
 PLS model for creamy mouthfeel, 107, 111
 PLS model for creamy mouthfeel of yogurts, 111
 PLS regression for creamy mouthfeel *versus* texture attributes, 113, 120–121
Peanuts, 206, 207
 See also Textured peanut (TP)
2-Pentylpyridine, 74
Phase partitioning
 air-product partition coefficient (P_{ap}), 146–147
 air-water partition coefficient (Kaw), 167, 168f
 concentration gradients of aroma compounds in water and air, 148–149
 concentration gradients of aroma compounds in water and emulsions, 150–151
 diffusion coefficient (D; Stokes-Einstein equation), 149–150
 dynamic headspace system model, 166–169
 effect of partition coefficient on mass transfer, 166–169
 effects of lipids on volatility (P_{ap}) of aroma compounds, 147f
 gas phase analysis methods, 160–161
 oil-water partition coefficient (P_{ow}), 62, 146–147
 oil-water partition coefficients, effect of chain length and saturation, 62, 148, 192
 temperature, effect on partition between lipid and vapor phases, 62, 148
 vapor-oil partition coefficient (K_i), effect on flavor release in emulsions, 63–64, 67, 68t, 69
Phenolics, 12
Phenylpropanoids, 12
Phosphatidic acid (PA), 20f, 31
Phosphatidylcholine (PC, lecithin)
 fatty acid composition, 28t
 formation of odorants from PC, 25–28
 model degradation reactions, 28–29
 molecular organization and self-assembly, 31–33
 odor activity values (OAV) of odorants from PC, 27
 odor quality, 23
 odorants in aroma extract of PC, 23–24, 25t
 phase diagrams, 32–33
 sources, 11
 structure, 20f
 uses, 11
 volatiles in heated PC, 23
 See also Phospholipids
Phosphatidylethanolamine (PE, cephalin)
 fatty acid composition, 28t
 formation of odorants from PE, 25–28
 model degradation reactions, 28–29
 molecular organization and self-assembly, 31–33

odor activity values (OAV) of odorants from PE, 27
odor quality, 23
odorants in aroma extract of PE, 23–24, 25t
phase diagrams, 32–33
structure, 20f
See also Phospholipids
Phosphatidylinositol (PI), 20f
Phospholipids
  antioxidant effects, 11
  fatty acid content, 20, 28t
  fatty acids, effect on odorant formation, 29–31
  model system for degradation, 21–22, 28–29
  molecular organization and self-assembly, 31–33
  phase diagrams, 32–33
  phosphatidylserine, 11
  polar moiety, effect on odorant formation, 31
  properties and structure, 11, 19–20
  sources, 19–20
  uses, 19
  volatile compounds in heated phospholipids, 20, 23–26
  See also Phosphatidylcholine; Phosphatidylcholine (PC, lecithin); Phosphatidylethanolamine; Phosphatidylethanolamine (PE, cephalin); Phosphatidylinositol
Phytosterols, 11
PLS regression for creamy mouthfeel *versus* texture attributes, 113
Polyunsaturated fatty acids (PUFA), 5–8
  See also n-3 fatty acids; n-6 fatty acids; specific compounds
Pork, 80
Potatoes, flavor, 50
  See also French fries
PUFA. *See* Polyunsaturated fatty acids (PUFA)
Pyrazines, 207

R

Ranciment induction period of oils, 8t
Rapeseed oil, 201, 202t
Rheology, terminology, 137–138
  See also Cheese chemistry and rheology
Romano cheese, changes during ripening, 138, 139f
Rough mouthfeel
  effect on creamy mouthfeel, 106, 113, 114–115, 121, 127–128, 130
  physiological and physical basis, 121
  time intensity measurements, 122–123, 130, 131f

S

Saliva
  effect of dilution by saliva or water on flavor release, 160, 163–165
  effect of starch breakdown by α-amylase creamy mouthfeel, 100–101, 121–122, 126–127, 128, 130
  role in lubrication, 96, 97, 100–101, 166
Saturated lipids, health effects, 5
Sauces, 99f, 100–101
Semi-solid foods, 88, 98t, 106
  See also Custards; Emulsions; Mayonnaise; Yogurts (yoghurts)
Sensory attributes
  evaluation by sensory panel, 99–100, 122–123
  relationship to friction, 100f, 101
  See also Creamy mouthfeel; Rough mouthfeel; Texture attributes
Sesame oil, 12
Sesamin, 12
Sesaminol, 12
Sesamol, 12
Shark liver oil, 12
Sheep meat
  2-acetyl-1-pyrroline, 79

2-aminoacetophenone, 79
Aroma Extraction DIlution Analysis (AEDA), 74, 75–78
branched chain fatty acids in, 74, 80–81
characteristic aroma, 73–74
comparison of adipose tissue and intramuscular fat, 81
comparison of odor-active oxo-compounds in different types of meat, 80
consumption, 73
(E,E)-2,4 decadienal, 76, 77, 78, 79–80, 81–82
(E)-2-decenal, 81–82
4,5-epoxy-(E)-2-decenal in, 76, 77, 78, 81–82
4-ethyloctanoic acid in, 74, 76, 77, 78, 79–80, 81
flavor dilution (FD) chromatogram of odorants from cooked sheep meat, 75, 76f
flavor dilution (FD) factors of odorants from raw and cooked sheep meat, 77–79, 78t
flavor dilution (FD) factors of odorants from raw sheep meat, 78t
GC/Olfactometry, 74
3-hydroxy-4,5-dimethyl-2(5H) furanone, 76
4-hydroxy-2,5-dimethyl-3(2H) furanone, 76, 77f, 79
4-methylnonanoic acid in, 74, 79–80
4-methyloctanoic acid in, 74, 81
(E,Z)-2,6-nonadienal, 76, 77f, 79f
(E)-2-nonenal, 76, 77f, 79–80, 81–82
(Z)-1,5-octadien-3-one, 76, 77
1-octen-3-one, 76, 77f
odorants in cooked lean sheep meat, 75–76, 77f, 79–80
odorants in raw lean sheep meat, 77–79, 77–80
odorants in sheep adipose tissue fat, 81–82
2-pentylpyridine in, 74

volatile compounds in cooked meat, 74
Solid-phase microextraction (SPME) measurement of aroma release, 63, 64, 65f–66f, 67
Sources of lipids and oils, 4
Soybean oil, 17t
Springiness, definition, 137
Squalene, 12–13
Stability of edible oils
  effect of degree of unsaturation, 7, 8t
  effect of fatty acid position in triacylglycerol (Sn-1, Sn-2, Sn-3), 7
  effect of processing, 13–18
  induction period of oils, 7, 8t
  iodine value (IV) of oils, 7
Stearic acid (C18:0) health effects, 5
Sterols, 17t
Stokes-Einstein equation, 149–150
Stress-strain curves, 137
Structure breakdown by α-amylase. See α-Amylase
Sunflower oil, 201, 202t

T

Taste, definition, 146
Taste perception, effect of lipids, 154
Texture, definition, 137
Texture attributes
  airy attribute, effect on creamy mouthfeel, 88
  creamy mouthfeel relationship to texture attributes, 88, 106, 127–128, 129f
  fatty attribute, effect on creamy mouthfeel, 88
  granularity or grittiness, effect on creamy mouthfeel, 106
  PLS regression for creamy mouthfeel versus texture attributes, 113
  pre- and post-swallow components, 88

rough mouthfeel, effect on creamy mouthfeel, 106
smooth attribute, effect on creamy mouthfeel, 106
thick attribute, effect on creamy mouthfeel, 106
Texture characteristics, categories, 88, 106
Texture profile analysis (TPA), 137
Textured peanut (TP)
  extrusion processing, 206, 208
  fatty acid profile, 209, 211–212
  heterocyclic compounds, 213–215
  lipid oxidation and lipid oxidation products, 211, 215–217
  moisture content, 209, 210–211, 212
  preparation, 208–209
  uses, 206
  volatile flavor compounds, 209–210, 213
  water activity (a_w), 211
Tocols, 8, 9
  See also Tocopherols; Tocotrienols
Tocopherols (T), 9, 17*t*
  See also Tocols
Tocotrienols (T3), 9
  See also Tocols
Torsion tests, 138
Transient tests, 138
Triacylglycerols (TAG)
  amounts in animal fat, 67, 68*t*, 69
  differential scanning calorimetry (DSC) thermograms, 193, 195, 196*f*
  effect of fatty acid chain length on volatility, 192, 196–197, 198*f*
  effect of fatty acid saturation on volatility, 192
  effect of temperature on aroma volatility, 196, 197*f*
  fatty acids composition, 5–7, 193
  fatty acids profile of commercial fats, 200*t*
  fatty acids profile of commercial oils, 201, 202*t*

gas chromatographic analysis, 194
in cheese, 134
linearization model, application to complex lipids and emulsions, 199–203
linearization model for carbon number, chromatographic surface area, and unsaturation, 197, 199, 203
melting temperatures, 195, 196*f*
predominance in edible fats and oils, 4, 5
tributyrin (C 4:0), 192, 195
tricaprylin (C 8:0), 192, 195
trilaurin (C 12:0), 192, 195
trimyristin (C 14:0), 192
triolein (C 18:1), 192, 195
tristearin (C 18:0), 192, 195
weighted average of carbon number and degree of unsaturation in commercial fats and oils, 200*t*, 202*t*, 203
(E,Z,Z)-2,4,7-Tridecatrienal, 24–27, 29

U

Ubiquinones (coenzyme Q), 9, 11
(E)-2-Undecenal, 24–27
Uniaxial compression, 137
Unsaponifiable matter, definition, 4
Unsaturation, 7, 8*t*

V

Vapor-oil partition coefficient (K_i), 63–64, 67, 68*t*, 69

X

Xanthophylls, 12

Y

Yogurts (yoghurts)
  composition of commercial yogurts, 107, 110*t*
  effect of lipids on flavor delivery, 161–162, 163*f*, 164
  evaluation by panel, 107
  PLS model for creamy mouthfeel, 111, 114, 115*f*